EULER'S GEM

Euler's Gem

The Polyhedron Formula and the Birth of Topology

David S. Richeson

With a New Preface by the Author

Princeton University Press
Princeton and Oxford

Copyright © 2008 by Princeton University Press

Preface to the Princeton Science Library edition,
copyright © 2019 by Princeton University Press

Published by Princeton University Press, 41 William Street,
Princeton, New Jersey 08540

In the United Kingdom: Princeton University Press, 6 Oxford Street, Woodstock,
Oxfordshire OX20, 1TR

press.princeton.edu

First printed in 2008
New Princeton Science Library edition, with a new preface by the author, 2019
Paperback ISBN 978-0-691-19137-9

The Library of Congress has cataloged the cloth edition of this book as follows

Richeson, David S.
Euler's gem : the polyhedron formula and the birth of topology / David S. Richeson.
p.   cm.
Includes bibliographical references and index.
ISBN-13: 978-0-691-12677-7 (alk. paper)
ISBN-10: 0-691-12677-1 (alk. paper)
1. Topology–History.   2. Polyhedra.   I. Title.
QA611.A3R53 2008
514.09–dc22   2008062108

British Library Cataloging-in-Publication Data is available
This book has been composed in Aldus
Printed on acid-free paper. ∞
Printed in the United States of America

To Ben and Nora
your faces
all your edges
I love you from vertex to toe

Contents

PREFACE TO THE PRINCETON SCIENCE LIBRARY EDITION

"Either this is madness or it is Hell."
"It is neither," calmly replied the voice of the Sphere, "It is
Knowledge; it is Three Dimensions: open your eye once again
and try to look steadily."
I looked, and, behold, a new world!
—Edwin Abbott, *Flatland*, 1884

When I was growing up, the general consensus among kids was that there were math people and there were writing people—the two groups were disjoint, separated by a vast chasm. I was good at and enjoyed mathematics; therefore, I was not good at writing. That's what I told myself. Despite doing well in my writing classes, this belief persisted for many years.

In graduate school, when I started writing about mathematics, I realized that I was good at writing and—gasp!—that I enjoyed it. I got great pleasure from taking a sprawling mass of facts, finding a way to organize them in a logical fashion, teasing out the big picture from the details, and presenting them in a way that made sense and was understandable to readers. I found it rewarding to go over my writing again and again, trying to streamline the prose and to read it as if I was reading it for the first time.

Also, when I was in graduate school, I was repeatedly amazed by Leonhard Euler's polyhedron formula and the Euler characteristic. The expression $V-E+F$ seemed to pop up everywhere. It was simultaneously elementary and deep, theoretically interesting and practically useful. It became an ever-present friend. I realized that because this mathematical gem typically showed up in advanced areas of mathematics, it was hidden from the general population and even from many mathematics students. At some point, I recognized that there should be a book about Euler's formula and that I should write it. I had never contemplated writing a book—it was not on my bucket list. But suddenly I was bitten by the bug. I started brainstorming ideas—listing all of the topics that I knew had connections to Euler's formula. I scoured journals and books in our library for any appearance of this formula.

I quickly realized that not only was the mathematics interesting, but so was the history. What did Euler do? Why does the formula sometimes have Descartes's or Poincaré's name attached to it? Why had the Greeks not discovered this formula? How did it transition from a theorem about geometry to one about topology? And who are these interesting individuals who contributed to the story? Although I had read some books on the history of mathematics, I had no training in this area. So, I immersed myself in the literature, and the more I read, the more I fell in love with the field. I even designed and taught several history of mathematics courses at my college.

One of the surprising challenges I faced was determining exactly what Euler's contribution was. I could find only vague statements that his proof was flawed and proofs by other mathematicians. Euler wrote his article in Latin, so I enlisted my friend and colleague at Dickinson College, the classicist Chris Francese, to help me translate it sentence by sentence. It was an amazing and eye-opening experience to read what, exactly, Euler wrote. His proof did indeed have a flaw, but it was one that could be easily fixed.

During this time, I was working on the book as a side project. Much of my time was spent teaching and conducting research in topology and dynamical systems with my longtime collaborator, Jim Wiseman. But, little by little, I made progress on the book. It took a while for me to decide who the audience should be. At first, I started writing it as if it would be a textbook— one with exercises at the end of each chapter. But I had always been a fan of the mathematics books published by Princeton University Press. They were aimed at a general audience, but they did not dumb down the math. Many of them were also written by mathematicians, not by journalists who might not have the same depth of knowledge about the subject. Eventually, I realized that I should model my book on those in the Princeton catalog.

Of course, my book covered very advanced topics—topology, dynamical systems, differential geometry, graph theory, and so on. What made me think that I could pitch it at a level that a general reader would be able to follow? In the front matter to *A Brief History of Time* (1988), Stephen Hawking wrote, "Someone told me that each equation I included in the book would halve the sales." And my book is *about* an equation!

In the end, I did not take Hawking's editor's implied advice. I decided that it was all right to show real math. I trusted my judgment and had confidence in the readers. I would not talk down to them. I would not omit the beautiful and interesting pieces of mathematics just because they were challenging to visualize or required equations. In my heart, I believed that the mathematics that a student might not see until college or graduate school was not that complicated—at least the ideas behind it were not that complicated. I took it as a challenge to find ways to present this mathematics so that my readers could follow it. I spent a lot of time drawing figures that I hoped would be helpful. I gave talks on this material to many different audiences, and I taught some of the material in my history of mathematics class. There's no substitute for the instant feedback of seeing your audience nodding along in understanding or staring blankly at an explanation you just gave.

Still, I was not expecting anyone to read the book. After years of writing research articles that were each read by a handful of experts in the field, I was stunned by the reception of the book—it received very positive reviews; it was awarded the Mathematical Association of America's Euler Book Prize, which is conferred on an "outstanding book about mathematics"; it was translated into other languages; and, most significant, people read it! I have enjoyed hearing from readers—whether they are mathematicians, students, or people who love mathematics but no longer do mathematics. Writing the book felt like a selfish pursuit in which I investigated the topics that interested me. But its success shows that these ideas interest others as well.

In hindsight, I shouldn't have been surprised that readers enjoyed the book—I was working with great material. People love topology—rubber sheet geometry, one-sided surfaces, higher dimensional spaces, and so on. They love polyhedra. And they love Euler's formula. Moreover, in recent years, topology has experienced a renaissance—it is a "cool" subject. In 2003, Grigori Perelman proved the Poincaré conjecture, one of the most famous problems in mathematics, and then he turned down both the Fields Medal and the million-dollar Clay Millennium Prize. Also, like a lot of mathematics, topology has gone from being purely an academic, theoretical discipline with no clear applications to one that is useful and surprisingly well-suited to computation. Today there are books, journals, and conferences on using topology to study networks, data analysis, qualitative solutions to problems, robotics, protein folding, digital imagery, and on and on.

Plus, it was a perfect time to write about Euler. We had just celebrated his three hundredth birthday in 2007, so there was already a buzz about this genius who was the equal of Isaac Newton, Karl Gauss, and Archimedes, yet somehow was not as well known. Finally, there was a gap in the literature; although many people have written about Euler's polyhedron formula, no one had written *this book*—a 2,500-year history from the ancient Greeks up to the present day.

I am grateful for all of the doors *Euler's Gem* opened for me. I have been invited to speak about my book, I have met many fascinating people, and it has provided me new professional opportunities. For instance, I was selected to be editor of *Math Horizons*, the undergraduate magazine of the Mathematical Association of America—a publication that strives to bring interesting mathematics and intriguing stories about mathematicians to a general audience. It was a perfect marriage of my love of mathematics and expository writing.

I would like to thank my editor at Princeton University Press, Vickie Kearn, for believing in me and believing in the project. She was very kind to me when I received a blunt reader's report that asserted the book should end with Euler's 1750 proof and that I should cut all of the material on topology. After discussing the critique, Vickie and I realized it was a problem of expectations: I needed to do a better job of letting the reader know the scope of the book at the outset. This required two small changes—adding a subtitle (*The Polyhedron Formula and the Birth of Topology*) and writing a preface that described the book in its entirety. It seemed to do the trick—none of the reviews of the book have made this same criticism.

It is a great honor to have *Euler's Gem* included in the Princeton Science Library series alongside books by Albert Einstein, Richard Feynman, George Polya, Hermann Weyl, Stephen Hawking, Roger Penrose, and other esteemed scientists and mathematicians. I truly wish I could go back to my childhood self and say: don't believe the lie; people can be both mathematicians and writers.

David Richeson
Carlisle, Pennsylvania
December 2018

PREFACE

A mathematician is a device for turning coffee into theorems.
—Alfréd Rényi, oft repeated by Paul Erdős[1]

In the spring of my senior year of college I told an acquaintance that I would be pursuing a PhD in mathematics in the fall. He asked me, "What will you do in graduate school, study really big numbers or calculate more digits of pi?"

It is my experience that the general public has little idea what mathematics is and certainly has no conception what a research mathematician studies. They are shocked to discover that new mathematics is still being created. They think that mathematics is only the study of numbers or that it is a string of courses that terminates at calculus.

The truth is, I have never been that interested in numbers. Mental arithmetic is not my strong point. I can split a dinner check and calculate a tip at a restaurant without reaching for a calculator, but it takes me about as long as it does anyone else. And calculus was my least favorite mathematics class in college.

I enjoy looking for patterns—the more visual the better—and untwisting intricate logical arguments. The shelves in my office are full of books of puzzles and brain teasers with my childhood pencil marks in the margins. Move three matchsticks to form this other pattern, find a path through this grid that satisfies this list of rules, cut up this shape and rearrange it to be square, add three lines to this picture to create nine triangles, and other mind-benders. To me, *this* is mathematics.

Because of my love of spatial, visual, and logical puzzles, I have always been attracted to geometry. But in my senior year of college I discovered the fascinating field of topology, generally understood to be the study of nonrigid shapes. The combination of beautiful abstract theory and concrete spatial manipulations fit my mathematical tendencies perfectly. The loose and flexible topological view of the world felt very comfortable. Geometry seemed straight laced and conservative in comparison. If geometry is dressed in a suit coat, topology dons jeans and a T-shirt.

This book is a history and celebration of topology. The story begins with its prehistory—the geometry of the Greek and Renaissance mathematicians and their study of polyhedra. It continues through the eighteenth and nineteenth centuries as scholars tried to come to grips with the idea of shape and how to classify objects without the rigid conditions imposed by geometry. The story culminates in the modern field of topology, which was developed in the early years of the twentieth century.

As students, we learned mathematics from textbooks. In textbooks, mathematics is presented in a rigorous and logical way: definition, theorem, proof, example. But it is not discovered that way. It took many years for a mathematical subject to be understood well enough that a cohesive textbook could be written. Mathematics is created through slow, incremental progress, large leaps, missteps, corrections, and connections. This book shows the exciting process of mathematical discovery in action—brilliant minds thinking about, questioning, refining, pushing, and altering the work of their predecessors.

Rather than giving a simple history of topology, I chose Euler's polyhedron formula as a tour guide. Discovered in 1750, Euler's formula marks the beginning of the transition period from geometry to topology. The book follows Euler's formula as it evolved from a curiosity into a deep and useful theorem.

Euler's formula is an ideal tour guide because it has access to marvelous rooms that are rarely seen by other visitors. By following Euler's formula we see some of the most intriguing areas of mathematics—geometry, combinatorics, graph theory, knot theory, differential geometry, dynamical systems, and topology. These are beautiful subjects that a typical student, even an undergraduate mathematics major, may never encounter.

Also, on this tour I have the pleasure of introducing the reader to some of history's greatest mathematicians: Pythagoras, Euclid, Kepler, Descartes, Euler, Cauchy, Gauss, Riemann, Poincaré, and many others—all of whom made important contributions to this subject and to mathematics in general.

This book has no formal prerequisites. The mathematics that a student learns in a typical high school mathematics sequence—algebra, trigonometry, geometry—is sufficient, but most of it is irrelevant to this discussion. The book is self-contained, so in the rare cases that I need to, I will remind the reader of facts from these mathematics courses.

Do not be misled, though—some of the ideas are quite sophisticated, abstract, and challenging to visualize. The reader should be willing to read through logical arguments and to think abstractly. Reading mathematics is not like reading a novel. The reader should be prepared to stop and ponder each sentence on its own, reread an argument, try to come up with other examples, carefully examine the figures in the text, search for the big picture, and use the index to look back at the exact meaning of technical terminology.

Of course, there is no homework and no final exam at the end of the book. There is no shame in skipping over the difficult parts. If a particularly thorny argument is too difficult to grasp, jump to the next topic. Doing so will not sabotage the rest of the book. The reader may want to fold over the corner of a challenging page and come back to it later.

It is my belief that the audience for this book is self-selecting. Anyone who *wants* to read it should be *able* to read it. The book is not for everyone, but those who would not understand and appreciate the mathematics are precisely those who would never pick it up in the first place.

I had the precious advantage that I was not writing a textbook. I made every effort to be honest and rigorous in my descriptions of the mathematics, but I had the liberty to gloss over pesky details that confuse more than they illuminate. This way I could write at a higher level and focus on ideas, intuition, and the big picture. By necessity I was only able to give a superficial treatment of the many fascinating ideas in this book. Anyone interested in reading more about these topics or in seeing how the missing details complete the picture should consult the list of suggested readings in appendix B.

While this book is accessible to a broad audience, I also wrote it for mathematicians. Although parts of this book overlap with other books, there is no single resource that contains all this information. There is an extensive bibliography at the end of the book that includes many of the original papers. It should aid scholars who would like to dig deeper into the subject matter.

The book is organized as follows. It begins with the pre-Eulerian view of polyhedra in chapters 2, 3, 4, 5, and 6. These chapters focus on the most famous class of polyhedra, the regular polyhedra. Chapters 7, 9, 10, 12, and 15 present Euler's polyhedron formula and its generalizations to other rigid polyhedral shapes. This discussion takes us up to the middle of the nineteenth century. Chapters 16, 17, 22, and 23 focus on the topological view of Euler's formula that emerged at the end of that century. They cover surfaces and higher-dimensional topological objects.

The book also contains numerous applications of Euler's formula. Chapter 8 contains elementary uses of Euler's polyhedron formula. Chapters 11, 13, and 14 focus on graph theory. Chapters 18, 19, 20, and 21 focus on surfaces, their relationship with Euler's formula, and their application to knot theory, dynamical systems, and geometry.

I hope the readers of this book enjoy reading it as much as I enjoyed writing it. For me, this project was a giant puzzle—an academic scavenger hunt. Finding the pieces and assembling them into a cohesive story was a challenge and a joy for me. I love my job.

Dave Richeson
Dickinson College
July 6, 2007

EULER'S GEM

INTRODUCTION

Philosophy is written in this grand book—I mean the
universe—which stands continually open to our gaze, but it
cannot be understood unless one first learns to comprehend
the language and interpret the characters in which it is written.
It is written in the language of mathematics, and its
characters are triangles, circles, and other geometrical figures,
without which it is humanly impossible to understand a single
word of it; without these one is wandering about in a
dark labyrinth.
—Galileo Galilei[1]

They all missed it. The ancient Greeks—mathematical luminaries such as Pythagoras, Theaetetus, Plato, Euclid, and Archimedes, who were infatuated with polyhedra—missed it. Johannes Kepler, the great astronomer, so in awe of the beauty of polyhedra that he based an early model of the solar system on them, missed it. In his investigation of polyhedra the mathematician and philosopher René Descartes was but a few logical steps away from discovering it, yet he too missed it. These mathematicians, and so many others, missed a relationship that is so simple that it can be explained to any schoolchild, yet is so fundamental that it is part of the fabric of modern mathematics.

The great Swiss mathematician Leonhard Euler (1707–1783)—whose surname is pronounced "oiler"—did not miss it. On November 14, 1750, in a letter to his friend, the number theorist Christian Goldbach (1690–1764), Euler wrote, "It astonishes me that these general properties of stereometry [solid geometry] have not, as far as I know, been noticed by anyone else."[2] In this letter Euler described his observation, and a year later he gave a proof. This observation is so basic and vital that it now bears the name *Euler's polyhedron formula.*

A polyhedron is a three-dimensional object such as those found in figure I.1. It is composed of flat polygonal *faces*. Each pair of adjacent faces meets along a line segment, called an *edge*, and adjacent edges meet at a corner, or a *vertex*. Euler observed that the numbers of vertices, edges,

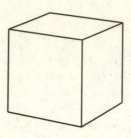 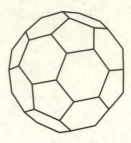

Figure I.1. A cube and a soccer ball (truncated icosahedron) both satisfy Euler's formula.

and faces (V, E, and F) always satisfy a simple and elegant arithmetic relationship:

$$V - E + F = 2.$$

The cube is probably the best-known polyhedron. A quick count shows that it has six faces: a square on the top, a square on the bottom, and four squares on the sides. The boundaries of these squares form the edges. Counting them, we find twelve total: four on the top, four on the bottom, and four vertical edges on the sides. The four top corners and the four bottom corners are the eight vertices of the cube. Thus for the cube, $V = 8$, $E = 12$, and $F = 6$, and of course

$$8 - 12 + 6 = 2, \qquad \checkmark$$

as claimed. It is more tedious to count them, but the soccer-ball-shaped polyhedron shown in figure I.1 has 32 faces (12 pentagons and 20 hexagons), 90 edges, and 60 vertices. Again,

$$60 - 90 + 32 = 2.$$

In addition to his work with polyhedra, Euler created the field of *analysis situs*, known today as topology. Geometry is the study of rigid objects. Geometers are interested in measuring quantities such as area, angle, volume, and length. Topology, which has inherited the popular moniker "rubber-sheet geometry," is the study of malleable shapes. The objects a topologist studies need not be rigid or geometric. Topologists are interested in determining connectedness, detecting holes, and investigating twistedness. When a carnival clown bends a balloon into the shape of a dog, the balloon is still the same topological entity, but geometrically it is very different. But when a child bursts the balloon with a pencil, leaving a gaping

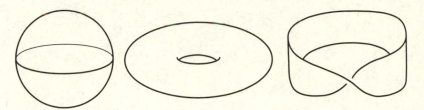

Figure I.2. Topological surfaces: a sphere, a torus, and a Möbius band.

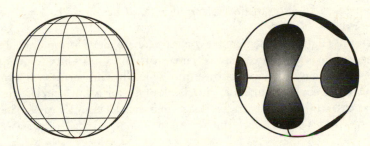

Figure I.3. Two partitions of the sphere.

hole in the rubber, it is no longer topologically the same. In figure I.2 we see three examples of topological surfaces—the sphere, the doughnut-shaped torus, and the twisted Möbius band.

Scholars in this young field of topology were fascinated by Euler's formula, and they attempted to apply it to topological surfaces. The obvious question arose: where are the vertices, edges, and faces on a topological surface? The topologists disregarded the rigid rules set forth by the geometers and allowed the faces and edges to be curved. In figure I.3 we see a partition of a sphere into "rectangular" and "triangular" regions. The partition is formed by drawing 12 lines of longitude, meeting at the two poles, and 7 lines of latitude. This globe has 72 curved rectangular faces and 24 curved triangular ones (the triangular faces are located near the north and south poles), giving a total of 96 faces. There are 180 edges and 86 vertices. Thus, as with polyhedra, we find that

$$V - E + F = 86 - 180 + 96 = 2.$$

Likewise, the 2006 World Cup soccer ball, which consists of six four-sided hourglass-shaped patches and eight misshapen hexagonal patches

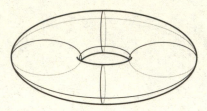

Figure I.4. A partition of the torus.

(see figure I.3), also satisfies Euler's formula (it has $V = 24$, $E = 36$, and $F = 14$).

At this point we are tempted to conjecture that Euler's formula applies to every topological surface. However, if we partition a torus into rectangular faces, as in figure I.4, we obtain a surprising result. This partition is formed by placing 2 circles around the central hole of the torus and 4 circles around its circular tube. The partition has 8 four-sided faces, 16 edges, and 8 vertices. Applying Euler's formula we find

$$V - E + F = 8 - 16 + 8 = 0,$$

rather than the expected 2.

If we were to construct a different partition of the torus we would find that the alternating sum is still zero. This gives us a *new* Euler's formula for the torus:

$$V - E + F = 0.$$

We can prove that every topological surface has its "own" Euler's formula. No matter whether we partition the surface of a sphere into 6 faces or 1,006 faces, when we apply Euler's formula we will always get 2. Likewise, if we apply Euler's formula to any partition of the torus, we will get 0. This special number can be used to distinguish surfaces just as the number of wheels can be used to distinguish highway vehicles. Every car has four wheels, every tractor trailer has eighteen wheels, and every motorcycle has two wheels. If a vehicle does not have four wheels, then it is not a car; if it does not have two wheels, then it is not a motorcycle. In the same way, if $V - E + F$ is not 0, then topologically the surface is not a torus.

The sum $V - E + F$ is a quantity intrinsically associated with the shape. In the lingo of topologists, we say that it is an *invariant* of the surface. Because of this powerful property of invariance, we call the number

 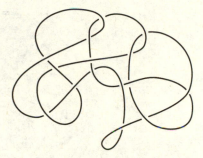

Figure I.5. Are these the same knot?

$V - E + F$ the *Euler number* of the surface. The Euler number of a sphere is 2 and the Euler number of a torus is 0.

At this point, the fact that every surface has its own Euler number may seem nothing more than a mathematical curiosity, an "isn't that cool" kind of fact to contemplate while holding a soccer ball or looking at a geodesic dome. This is most certainly not the case. As we will see, the Euler number is an indispensable tool in the study of polyhedra, not to mention topology, geometry, graph theory, and dynamical systems, and it has some very elegant and unexpected applications.

A mathematical knot is like an intertwined loop of string as shown in figure I.5. Two knots are the same if one knot can be deformed into the other without cutting and re-gluing the string. Just as we can use the Euler number to help distinguish two surfaces, with a little ingenuity we can also use it to distinguish knots. We can use the Euler number to prove that the two knots in figure I.5 are not the same.

In figure I.6 we see a snapshot of wind patterns on the surface of the earth. In this example there is a point off the coast of Chile where the wind is not blowing. It is located in the calm spot within the eye of the storm that is spinning clockwise. We can prove that there is always at least one point on the surface of the earth where there is no wind. This follows not from an understanding of meteorology, but from an understanding of topology. The existence of this point of calm comes from a theorem that mathematicians refer to as the hairy ball theorem. If we think of the wind directions as strands of hair on the surface of the earth, then there must be some point where the hair forms a cowlick. Colloquially, we say that "you can't comb the hair on a coconut." In chapter 19 we will see how the Euler number enables us to establish this bold assertion.

Figure I.6. Is there always a windless location on earth?

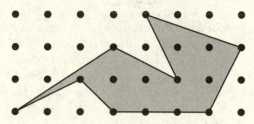

Figure I.7. Is it possible to determine the area of the shaded polygon by counting dots?

In figure I.7 we see a polygon sitting in an array of dots spaced one unit apart. The vertices of the polygon are located at the dots. Surprisingly, we can compute the precise area of the polygon simply by counting dots. In chapter 13 we will use the Euler number to derive the following elegant formula that gives the area of the polygon in terms of the number of dots that lie along the boundary of the polygon (B) and the number of dots in the interior of the polygon (I):

$$\text{Area} = I + B/2 - 1.$$

From this formula we conclude that our polygon's area is $5 + 10/2 - 1 = 9$.

There is an old and interesting problem that asks how many colors are required to color a map in such a way that every pair of regions with a common border are not the same color. Take a blank map of the United States and color it with as few crayons as possible. You will quickly discover that most of the country can be colored using only three crayons, but that a

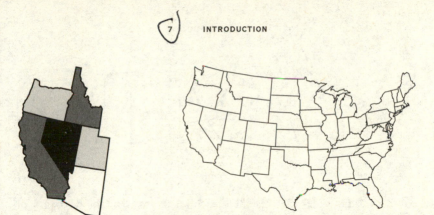

Figure I.8. Can we color the map of the United States using only four colors?

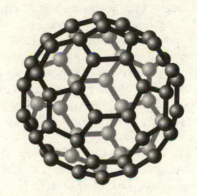

Figure I.9. The C_{60} buckminsterfullerene molecule.

fourth is needed to complete the map. For instance, since an odd number of states surround Nevada, you will need three crayons to color them—then you will need that fourth crayon for Nevada itself (figure I.8). If we are clever, then we can finish the coloring without using a fifth—four colors are sufficient for the entire map of the United States. It was long conjectured that every map can be colored with four or fewer colors. This infamous and very slippery conjecture became known as the four color problem. In chapter 14 we will recount its fascinating history, one that ended with a controversial proof in 1976 in which the Euler number played a key role.

Graphite and diamond are two materials whose chemical makeup consists entirely of carbon atoms. In 1985 three scientists—Robert Curl Jr., Richard Smalley, and Harold Kroto—shocked the scientific community by discovering a new class of all-carbon molecules. They called these molecules *fullerenes*, after the architect Buckminster Fuller, inventor of the geodesic dome (figure I.9). They chose this name because fullerenes are large

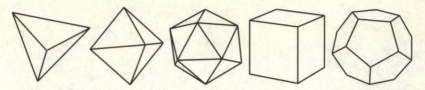

Figure I.10. The five regular solids.

polyhedral molecules that resemble such structures. For the discovery of fullerenes the three men were awarded the 1996 Nobel Prize in chemistry. In a fullerene every carbon atom bonds with exactly three of its neighbors, and rings of carbon atoms form pentagons and hexagons. Initially Curl, Smalley, and Kroto found fullerenes possessing 60 and 70 carbon atoms, but other fullerenes were discovered later. The most plentiful fullerene is the soccer-ball-shaped molecule C_{60} that they called the buckminster-fullerene. Remarkably, knowing no chemistry and only Euler's formula, we are able to conclude that there are certain configurations of carbon atoms that are impossible in a fullerene. For instance, every fullerene, regardless of the size, must have exactly 12 pentagonal carbon rings even though the number of hexagonal rings can vary.

For thousands of years people have been drawn to the beautiful and alluring regular solids—polyhedra whose faces are identical regular polygons (figure I.10). The Greeks discovered these objects, Plato incorporated them into his atomic theory, and Kepler based an early model of the solar system on them. Part of the mystery surrounding these five polyhedra is that they are so few in number—no polyhedron other than these five satisfies the strict requirements of regularity. One of the most elegant applications of Euler's formula is a very short proof that guarantees that only five regular solids exist.

Despite the importance and beauty of Euler's formula, it is virtually unknown to the general public. It is not found in the standard curriculum taught in the schools. Some high school students may know Euler's formula, but most students of mathematics do not encounter this relation until college.

Mathematical fame is a curious thing. Some theorems are well known because they are drilled into the heads of young students: the Pythagorean theorem, the quadratic formula, the fundamental theorem of calculus. Other results are thrust into the spotlight because they resolve a famous unsolved problem. Fermat's last theorem remained unproved for over three hundred years until Andrew Wiles surprised the world with his proof in

1993. The four color problem was posed in 1853 and was only proved by Kenneth Appel and Wolfgang Haken in 1976. The famous Poincaré conjecture was posed in 1904 and was one of the Clay Mathematics Institute Millennium Problems—a collection of seven problems deemed so important that the mathematician who solves one receives $1 million. The money is likely to be awarded to Grisha Perelman, who gave a proof of the Poincaré conjecture in 2002. Other mathematical facts are well known because of their cross-disciplinary appeal (the Fibonacci sequence in nature) or their historical significance (the infinitude of primes, the irrationality of π).

Euler's formula should be as well known as these great theorems. It has a colorful history, and many of the world's greatest mathematicians contributed to the theory. It is a deep theorem, and one's appreciation for this depth grows with one's mathematical sophistication.

This is the story of Euler's beautiful theorem. We will trace its history and show how it formed a bridge from the polyhedra of the Greeks to the modern field of topology. We will present many surprising guises of Euler's formula in geometry, topology, and dynamical systems. We will also give examples of theorems whose proofs rely on Euler's formula. We will see why this long-unnoticed formula became one of the most beloved theorems in mathematics.

CHAPTER 1

LEONHARD EULER AND HIS
THREE "GREAT" FRIENDS

Read Euler, read Euler, he the master of us all.
—Pierre-Simon Laplace[1]

We have become accustomed to hyperbole. Television commercials, billboards, sportscasters, and popular musicians regularly throw around sensational words such as greatest, best, brightest, fastest, and shiniest. Such words have lost their literal meaning—they are employed in the normal process of selling a product or entertaining a viewer. So, when we say that Leonhard Euler was one of the most influential and prolific mathematicians the world has ever seen, the reader's eyes may glaze over. We are not overselling the truth. Euler is widely considered, along with Archimedes (287–211 BCE), Isaac Newton (1643–1727), and Carl Friedrich Gauss (1777–1855), to be one of the top ten—or top five—most important and significant mathematicians in history.

In his life of seventy-six years, Euler created enough mathematics to fill seventy-four substantial volumes, the most total pages of any mathematician. By the time all of his work had been published (and new material continued to appear for seventy-nine years after his death) it amounted to a staggering 866 items, including articles and books on the most cutting-edge topics, elementary textbooks, books for the nonscientist, and technical manuals. These figures do not account for the projected fifteen volumes of correspondence and notebooks that are still being compiled.

Euler's importance is due not to his voluminous output but to the deep and groundbreaking contributions he made to mathematics. Euler did not specialize in one particular area. He was one of the great generalists: he had expertise that spanned the disciplines. He published influential articles and books in analysis, number theory, complex analysis, calculus, calculus of variations, differential equations, probability, and topology. This list does not include his contributions to such applied subjects as optics, electricity

Figure 1.1. Leonhard Euler.

and magnetism, mechanics, hydrodynamics, and astronomy. Furthermore, Euler possessed a trait that was and is rare among the top scholars: he was a first-rate expositor. Unlike the mathematicians who preceded him, Euler wrote using clear, simple language that made his work accessible to experts and students alike.

Euler was a gentle, unpretentious man whose life was centered around his large family and his work. He lived in Switzerland, Russia, Prussia, and then again Russia, and corresponded extensively with many important thinkers of the eighteenth century. His professional life was linked to three "Great" rulers of Europe—Peter the Great, Frederick the Great, and Catherine the Great. The legacies of these leaders include the creation or revitalization of their countries' national academies of science. These academies supported Euler so that he could spend time on pure research. The only repayment they expected was the occasional use of his scientific expertise for matters of the state and the recognition that his celebrity brought to the nation.

Leonhard Euler was born in Basel, Switzerland, on April 15, 1707, to Paul Euler and Marguerite Brucker Euler. Shortly afterward, the family moved to the nearby town of Riehen, where Paul took a job as the minister of the local Calvinist church.

Leonhard's earliest mathematical training was provided by his father. Although Paul was not a mathematician, he had studied mathematics under the famed Jacob Bernoulli (1654–1705). This instruction took place while Paul and Jacob's younger brother Johann (1667–1748), both students at University of Basel, boarded at Jacob's home. Jacob and Johann Bernoulli were members of what was to become the most esteemed family in mathematics. For over a century the Bernoulli clan played an important role in the advancement of mathematics, with at least eight Bernoullis making lasting contributions.

Leonhard began his formal studies at the University of Basel at the age of fourteen. This was not an unusually young age for a university student at that time. The university was quite small—it had only a few hundred students and nineteen professors. Paul hoped that his son would follow a career path into the ministry, so Leonhard studied theology and Hebrew. But his mathematical abilities were undeniable, and he quickly attracted the attention of his father's friend Johann Bernoulli. By this time Johann was one of the leading mathematicians in Europe.

Johann was an arrogant, brusque man with a competitive streak that produced storied rivalries (with, among others, his brother and one son). Yet he recognized the boy's remarkable talents and encouraged him to pursue mathematics. Euler wrote in his autobiography, "If I came across some obstacle or difficulty, I was given permission to visit him freely every Saturday afternoon and he kindly explained to me everything I could not understand."[2] These lessons played a valuable role in the maturation of Euler's mathematical skills.

Even as Leonhard excelled in his private mathematical studies, Paul held out hope that his son would enter the ministry. At the age of seventeen, Euler earned his master's degree in philosophy. Johann feared that mathematics might lose his protégé to the Church, so he intervened and told Paul in no uncertain terms that Leonhard had the potential to become a great mathematician. Because of his fondness for mathematics, Paul relented. Even though Euler abandoned the ministry, he remained a devout Calvinist for his entire life.

Euler's first independent mathematical accomplishment came at the age of nineteen. His theoretical work on the ideal placement of masts on a ship secured him an *accessit*, or "honorable mention," in a prestigious competition sponsored by the French Académie des Sciences. This feat would be incredible for any teenager, but was especially so for a Swiss youth who had never seen a ship on the ocean. Euler did not win the top prize for this competition, which would have been roughly equivalent to winning a Nobel Prize today, but in the years to come he won the highest honor on twelve occasions.

Figure 1.2. Peter the Great of Russia.

At the time of Euler's birth, a thousand miles to the northeast of Basel, the Russian tsar Peter the Great (1672–1725) was building the city of St. Petersburg. It was founded in 1703 on the marshy swampland where the Neva River flows into the Baltic Sea. Peter used forced labor to construct both the city and the strategically located Peter and Paul Fortress on an island in the Neva. He loved this new city, calling it his "paradise" and naming it after his patron saint. Despite the fact that most Russians, especially government officials, did not share Peter's feelings about this cold, wet place, he moved the Russian capital from Moscow to St. Petersburg. The young Euler had no way of knowing that this city would be his home for much of his life.

Peter the Great, a physically imposing figure who stood nearly seven feet tall, was the energetic, self-taught, determined leader of Russia from 1682 to 1725. Known as a ruthless reformer, he began the transformation of his country from an agrarian and feudal nation dominated by the Church into a powerful empire. His goal of modernizing—that is to say, Westernizing—Russian government, culture, education, military, and society was largely realized. As one Russian historian wrote, "All of a sudden, skipping entire epochs of scholasticism, Renaissance, and Reformation, Russia moved from a parochial, ecclesiastical, quasi-medieval civilization to the Age of Reason."[3]

As part of the process of Westernization, Peter wanted to reform Russia's educational system, which was nonexistent before his reign except for minimal teaching by the powerful Orthodox Church. As a result, Russia had no scientists. Because of the strong presence of the Church, Russians were fearful of scientific explanations of the world, preferring instead the traditional religious explanations. Peter recognized the need to improve Russia's international image and to dispel the notion that Russians hated science. He also knew that having a science program was crucial for creating and maintaining a powerful state.

Peter visited the Royal Society of London and the Académie des Sciences in Paris, both of which were founded in 1660. He was impressed with what he saw. He also admired the new Berlin Academy of Sciences, which was founded in 1700 upon the advice of Gottfried Leibniz (1646–1716). Leibniz is the famed mathematician who, along with Isaac Newton, is credited with the invention of calculus. These academies were not universities; they were "dedicated to the search for new knowledge and not the dissemination of existing wisdom."[4] The members of the academies were scholars, not teachers; their prime objective was the advancement of knowledge.

Peter wanted to create an academy such as the ones in Paris, London, and Berlin, and he wanted to establish it in his new city of St. Petersburg. For advice he turned to Leibniz. For nearly two decades, Peter and Leibniz had extensive conversations, both through letters and in face-to-face meetings, about education reform and the creation of an academy of science.

In 1724 Peter finalized his plans for the creation of the Academy of Sciences in St. Petersburg; it was the final and most ambitious project in his quest to improve Russia's educational system. However, he could not model his academy exactly on the European academies. Because Russia had no native scientists, he would have to persuade talented foreign scientists to relocate to St. Petersburg. Also, because Russia had no university system, the Academy of Sciences must function as a university. Part of the mandate of the Academy was to train Russians in science so that the Academy would not always have to rely on the foreigners.

Peter never saw the fruits of his labor; he died in early 1725. Thanks to the new empress, Peter's second wife, Catherine I (1684–1727), plans for the Academy continued. Foreign scholars began arriving within months of Peter's death, and the Academy of Sciences held its first meeting before the end of the year. Peter was fortunate that Catherine embraced the idea of the Academy. In the years that followed, the Academy was not always blessed with such sympathetic leaders. During the thirty-seven years between Peter's death and the coronation of Catherine the Great

(1729–1796), Russia was led by six rulers, and the Academy was always at the mercy of these opinionated and powerful people.

Initially the Academy was staffed by sixteen scientists: thirteen German, two Swiss, one French—and no Russians. The large German presence and the absence of Russians would later be a source of tension.

Because of the cold climate, the remote location, and the academic isolation, it was necessary to offer high salaries and provide comfortable accommodations to lure these scientists to St. Petersburg. The new academy was small, but it quickly fulfilled its promise of being an important, internationally renowned scientific institution. Eventually it became the center of all scientific scholarship in Russia. The Academy of Science has had several name changes, but it is still exists today and is known as the Russian Academy of Sciences.

Two of the foreign scholars who were stars of this new institution were Euler's friends, and Johann Bernoulli's sons, Nicolaus (1695–1726) and Daniel (1700–1782) Bernoulli. The two brothers had spoken to Euler about the Academy before they left Switzerland, and they promised to secure him a position as soon as possible. Immediately upon their arrival in Russia, they began lobbying the administrators of the Academy to hire their bright young friend. Their campaigning paid off quickly. In 1726 Euler was offered a position in the medical and physiology division. Unfortunately, Euler could not fully enjoy nor celebrate this exciting job offer. He was hired to fill an opening created by the tragic and untimely death of Nicolaus.

Euler was grateful for the job, but he did not move to Russia immediately. He had two reasons for remaining in Basel and putting his new job on hold. First, he had accepted a job in medicine, but he possessed only minimal knowledge in that area. So he decided to remain at the University of Basel and study anatomy and physiology. Second, he was stalling for time as he waited to hear if he would be offered a faculty position in physics at the University of Basel. In the spring of 1727, when he heard that he was not chosen for the job, he left for Russia. So began his life in St. Petersburg, where he lived for the next fourteen years and then again for the last seventeen years of his life.

Euler's journey to St. Petersburg by boat, foot, and wagon took seven weeks. On the day that he set foot in Russia, Empress Catherine I died after ruling for only two years. The fate of the new Academy was uncertain. Those who ran the country on behalf of the eleven-year-old tsar Peter II (1715–1730), the grandson of Peter the Great, saw the Academy as an expendable luxury and contemplated closing its doors. Fortunately, the school remained open, and in the confusion that ensued, Euler ended up where he rightly belonged—in the mathematical-physical division, not in the medical division. This first year of Euler's mathematical career, 1727, was also the year the mathematical giant Isaac Newton died.

Life at the Academy was difficult under Peter II, so the members of the Academy hoped that their fortunes would improve following the death of the fifteen-year-old tsar in 1730. The Academy did fare slightly better during the ten-year reign of Anna Ivanovna (1693–1740), but conditions in Russia turned bleak. Anna brought into her government a strong German influence, most notably her lover Earnst-Johann Biron (1690–1772). Biron was a ruthless tyrant who executed several thousand Russians and exiled tens of thousands more to Siberia. Those targeted by Biron included common criminals, Old Believers (of the Russian Orthodox Church), and Anna's political opponents. Later, when in Berlin, Euler was asked by the Queen Mother of Prussia why he was so taciturn. He replied, "Madam, it is because I have just come from a country where every person who speaks is hanged."[5]

In 1733, having had enough of the difficult Russian lifestyle and the internal politics at the Academy, Daniel Bernoulli moved back to Switzerland, and, at the age of twenty-six, Euler stepped into Daniel's role as head mathematician.

At this point, Euler realized that he might be in Russia for a long time, perhaps for the rest of his life. With the exception of hardships imposed by the political climate in Russia, Euler found a comfortable life. He acquired a good command of the Russian language, and he felt financially secure thanks to the higher salary that came with his promotion. So, in 1733 he decided to marry Katharina Gsell, the daughter of the Swiss-born painter Georg Gsell who had been brought to Russia by Peter the Great. Leonhard and Katharina started a family, eventually producing thirteen children. As was not uncommon in that day, only five of them lived past childhood, and just three outlived their parents.

Being a husband and a father did not slow down Euler's stream of publications. Now, and in every period of his professional life, he was an extremely active researcher. It is difficult to overstate Euler's massive output. Mathematical folklore says that he could write mathematics papers while bouncing a baby on his knee, and that he could compose a treatise between the first and second calls for dinner. He wrote about anything and everything. He produced masterpieces, short notes, corrections, explanations, partial results, ideas of proofs, introductory texts, and technical books.

No setback was able to slow Euler down. Even blindness could not impede the flood of his mathematical output. In 1738 he became ill after spending three long days working on an astronomical challenge. Although modern medicine casts this claim into doubt, it was long believed that this illness caused the deterioration and eventual loss of vision in his right eye. Euler took the loss of vision in stride. In his typically modest way he remarked, "Now I will have fewer distractions."[6] He later lost the vision in his other eye, and he lived in near-total darkness for the last seventeen years of his life. Despite his loss of sight he continued to make important mathematical contributions until the day of his death.

Euler's brain seemed hard-wired for mathematics in a way that other brains are not. He was able to juggle many abstract notions in his head simultaneously, and he was able to perform staggering mental calculations. In one famous story, two of Euler's students were adding seventeen fractional terms only to discover a discrepancy in their sums. Euler performed the sum in his head and settled the dispute by providing the correct answer. As the mathematician François Arago (1786–1853) famously wrote, "Euler calculated without apparent effort, as men breathe, or as eagles sustain themselves in the wind."[7] Modestly, Euler stated that his ability to manipulate symbols substituted for cleverness, saying that his pencil surpassed himself in intelligence.

Euler was also blessed with an amazing memory. He memorized count-less poems; from the time he was a child until his old age he was able to recite the entire text of Virgil's *Aeneid*, being able to recall the first and last sentences on any page. A more mathematical example of his prodigious memory was his ability to give the first six powers of the first hundred numbers. To put this in perspective, the sixth power of 99 is 941,480,149,401.

During his stay in St. Petersburg Euler devoted some of his time to projects for the state. In 1735 he was appointed the director of the Acad-emy's geography section and subsequently made important contributions to the creation of a much-needed map of Russia. He also wrote a two-volume book on shipbuilding that was so valuable that the Academy doubled his salary for that year.

Even while Euler enjoyed remarkable productivity, a happy family life, and a sizable income, conditions in Russia continued to worsen. The atmosphere at the Academy had become very tense, even hostile. Most of the senior faculty were German, and there was still very little Russian involvement. In the first sixteen years of the Academy's existence, only one Russian was given membership, and he was an adjunct who was never promoted to professor. The Russians resented the power possessed by the Germans and openly voiced anti-German opinions. Fortunately, the calm and reserved Euler was able to remain neutral in the internal politics at the Academy, but it made his working life stressful.

With the presence of Biron and the "German party" in Anna's government, fear and hatred of Germans continued to rise among the Russian people. In late 1740, shortly before her death, Anna appointed Biron regent for her successor, the two-month-old Ivan VI (1740–1764). After Anna's death the Russians' animosity toward the Germans came to a head— within a month Biron was overthrown, and a year later Ivan and the entire "German party" were removed from power. Peter the Great's daughter Elizabeth I (1709–1762) became the next empress.

During this period, life in Russia was dangerous, especially for non-Russians. Foreign academics were viewed suspiciously as possible spies for the West. Euler reacted to these conditions by staying quiet and by devoting all of his time to his work and family. In 1741 Euler was unable to tolerate life in Russia any longer, so he decided to leave St. Petersburg for Berlin.

The Berlin Academy of Sciences was founded in 1700 with the name Societas Regia Scientiarum. Leibniz had grand designs for the Academy. Like those in Paris and London, the Berlin Academy focused on science and mathematics, but unlike the others, it broadened its scope to include history, philosophy, language, and literature.

Despite Leibniz's high expectations, the Berlin Academy was slow to take off. Its difficulties were due in part to continual underfunding and to the internal French-German tensions. Conditions worsened following the accession of Frederick William I (1688–1740) in 1713. Under this anti-intellectual ruler, the Academy was completely neglected. The Berlin Academy showed none of the success of those in Paris and London. It was not a significant factor in the advancement of scientific knowledge; in fact, it was given the title "the anonymous society."

When Frederick William I died in 1740, his son Frederick II (1712–1786), later known as Frederick the Great, came to power. Although Frederick William I deliberately groomed his son for leadership, Frederick was in many ways his father's opposite. Tensions between the two ran deep. When he was eighteen Frederick was caught attempting to flee the country. His father forced Frederick to witness the execution of his friend and coconspirator (and, some say, homosexual lover).

Figure 1.3. Frederick the Great of Prussia.

Frederick was determined to extend the German territories, but he was also artistically and philosophically inclined. He aspired to be an enlightened ruler-philosopher. The revival of the Academy played an important role in his plan for revitalizing his country.

Unlike his father, Frederick disdained German culture and loved all things French. He changed the official name of the Berlin Academy to Académie Royale des Sciences et Belles Lettres. He insisted that French be the official language of the Academy, and he demanded that every article published in its journal be written in, or translated into, French. He preferred to be in the company of witty Frenchmen rather than calm and unemotional Germans. Voltaire (1694–1778) was one of his favorite correspondents and one of his closest advisors for matters relating to the Academy. It was Voltaire who first suggested to Frederick that he entice Euler to leave Russia and join the Berlin Academy.

Frederick had an intense aversion to the mathematical arts. In 1738 he wrote to Voltaire, "As for mathematics, I confess to you that I dislike it; it dries up the mind. We Germans have it only too dry; it is a sterile field which must be cultivated and watered constantly, that it may produce."[8] He viewed mathematics—and the sciences in general—as servants of the state. He judged the success of his scientists by their usefulness in practical matters. The scientists at the Academy were free to pursue their own projects so long as they attended to the requests made by the king.

At this time Euler was the most distinguished scholar in St. Petersburg, with a reputation that spread across Europe. Frederick set out to woo Euler. Despite the fact that Euler was troubled by the dangerous conditions in Russia, it took repeated contacts by Frederick to convince the Swiss mathematician to abandon St. Petersburg. In 1741 Euler assented, and by citing his declining health and the need for a warmer climate, he was able to take his leave of St. Petersburg.

At first Euler was satisfied in Berlin; he wrote to a friend in 1746, "The king calls me his professor, I think I am the happiest man in the world."[9] Unfortunately, this contentment did not last. In many ways life in Berlin was better than life in Russia, but Euler's experience was soured by Frederick's peculiar and surprising disdain for him. He referred to Euler as his "mathematical cyclops," ungraciously alluding to Euler's one good eye. Frederick's coolness was due in part to his dislike of mathematics, but that was not all. Euler's low-key and quiet demeanor did not sit well with Frederick, who viewed him as a simpleton. Frederick preferred the company of the witty, sophisticated, boisterous Voltaire. Also, Euler was a devout Calvinist. He read scripture to his family every evening, often accompanied by a sermon. Publicly Frederick espoused tolerance of religion, but privately he was a deist and had little respect for the pious Euler or his deeply held spiritual beliefs.

Euler harbored hard feelings toward Frederick as well. His greatest frustration while in Berlin was Frederick's refusal to make him president of the Academy. For several busy years, while he was fighting the Seven Years' War, Frederick was unable to find a suitable person to fill this position. During the interim Euler held the unofficial role of "acting president," but time and again Frederick passed him over as the permanent replacement. Euler performed well as interim president, but because he was not a philosopher capable of sharp, lively conversation, he would never gain Frederick's favor. The ultimate insult occurred in 1763 when Frederick admitted that he could not find a suitable replacement and declared himself president of the Academy.

There developed a further animosity between Euler and Frederick when in 1763, the king refused to allow one of Euler's daughters to marry a soldier because of the soldier's low rank. Perhaps the last straw was a series of heated exchanges between Frederick and Euler from 1763 to 1765. At issue was the sale of the state calenders (almanacs). These were made at great expense by the members of the Academy and were sold to the public to fund its operations. It was discovered that the chief commissioner had been pocketing funds from the sale of the calendars. Frederick and Euler disagreed over how to handle the corruption and mismanagement of this fundraising endeavor. It ended with Frederick sending a sharp rebuke to Euler.

While living in Berlin, Euler maintained good relations with his former colleagues in St. Petersburg. He remained editor for their journal, and he sent a total of 109 of his articles to be published in it. He tutored Russian students who were sent to Berlin. In exchange for the editing and tutoring, he was paid a regular stipend by the Russians. A more remarkable example of the Russians' respect for Euler occurred during the Seven Years' War. In the march on Brandenburg in 1760 the Russian army entered Carlottenburg. They came upon and pillaged a farm that belonged to Euler. When this action was discovered, the Russians—first the general, then Empress Elizabeth—paid Euler reparations that far exceeded the cost of the damages.

During the twenty-four years that Euler was in Berlin, the Russians were eager to get him to return to St. Petersburg. They approached him in 1746, in 1750, and in 1763, making generous offers to lure him back. Each time he refused, but he never closed the door completely. Finally in 1765, fed up with Frederick's hostility and seeing improved political conditions in Russia, he decided to return to St. Petersburg.

Despite his personal feelings, Frederick recognized Euler's prominent place in the international scientific community. Euler had published over two hundred works during his stay in Berlin. In 1749 he had been elected a Fellow of the Royal Society of London. In 1755 he had been appointed the ninth foreign member of Académie des Sciences in Paris, even though the number of foreign members was limited to eight. He had also served the state well; in addition to the creation of the calendars, Euler worked on coinage for the national mint, the placement of canals, the design of aqueducts, the creation of pensions, and the improvement of artillery.

Frederick attempted to prevent Euler's departure. Euler was forced to make repeated requests for permission to leave. In 1766 Frederick finally relented and allowed Euler to depart; at the age of 59, Euler and his eighteen dependents returned to St. Petersburg.

Later that year, on the recommendation of the French mathematician Jean D'Alembert (1717–1783), Frederick replaced Euler with Joseph-Louis Lagrange (1736–1813), a rising star who turned out to be a stellar mathematician. In his typically caustic way, the king wrote to D'Alembert thanking him "for having replaced a half-blind mathematician by a mathematician with both eyes, which will especially please the anatomical members of the academy."[10] Ironically, despite Frederick's distaste for

Figure 1.4. Catherine the Great of Russia.

mathematics and his love of philosophy, his Academy will forever be remembered for its impressive roster of mathematicians, and not for its philosophers.

At the end of his stay in Berlin, while Euler was clashing with Frederick, Russia was under the rule of Peter III (1728–1762), a miserable, psychologically unstable, pro-German leader who was known to "fear and despise Russia and the Russians."[11] In 1762 his rule came to an abrupt end when he was overthrown by his wife, who then took the throne as Catherine II. Shortly thereafter, perhaps under Catherine's orders, Peter was murdered by guards holding him in custody.

Catherine, who later became known as Catherine the Great, was empress until 1796. Just as the eighteenth century began under the rule of the powerful and influential Peter the Great, so was the century's end marked by the distinguished leadership of Catherine the Great. She was an intelligent, strong-willed, ambitious, and energetic leader. As the French philosopher Denis Diderot (1713–1783) said after visiting Catherine's court, she "is the soul of Caesar with all the seductions of Cleopatra."[12] Under her rule the quality of life in Russia showed a marked improvement. Education, which had largely been ignored since the time of Peter the Great, was again a priority for the Russian government.

During the early days of the Academy, the institution shone due to Euler's brilliance. When he moved to Berlin, so did the mathematical spotlight. This loss, together with the years of rapid political turnover, made it difficult for the institution to attract talented foreign scholars. The Academy was on shaky ground. One of Catherine's projects for educational reform was to revive the St. Petersburg Academy and elevate it to its earlier heights. As the mathematician André Weil (1906–1998) wrote, "This was almost synonymous with bringing Euler back."[13]

Catherine saw to it that Euler's stiff demands were met and exceeded. He received double the salary that was offered to him in 1763, his wife received a stipend, his eldest son was hired by the Academy, and his younger sons were guaranteed future employment. In addition, Catherine provided Euler a fully furnished house, and he was given one of her own cooks. Upon his arrival in St. Petersburg, Euler was greeted warmly by the Empress. His return to the Academy refocused the attention of the mathematical community on St. Petersburg and ensured the continued success of the Academy.

There are similarities between Catherine the Great and Frederick the Great; they were, after all, prime examples of "enlightened despots." However, Euler's relationships with the two leaders were very different. His experience in Catherine's St. Petersburg was much more positive than was his experience in Frederick's Berlin. Catherine was a lover of science, and she welcomed Euler as a celebrity. He was the ranking academician with the most administrative power of any scholar.

In his lifetime, Euler saw many changes in the capital city of St. Petersburg. The city was only twenty-four years old when he first arrived, sixty-three years old when he returned, and eighty years old when he died. By the end of the eighteenth century the population had grown to more than 166,000. St. Petersburg was home to some of the wealthiest noblemen in the empire and some of the poorest peasants. Nearly one quarter of the population was military.[14] St. Petersburg continued to be loved by some Russians and hated by others (this is still true today). In keeping with Peter the Great's plan, the city was full of beautiful European-style architecture. It was the most European of all Russian cities. It came to be known as the "Venice of the north" because of its many islands and waterways.

Euler's second stay in St. Petersburg was a time of professional success, but it was peppered with episodes of personal loss. In 1771 his house burned to the ground. The quick action of a selfless servant who carried him out of the burning building saved Euler's life. His entire library was destroyed by the fire, but thankfully for science, his manuscripts were rescued. In response to the tragedy Catherine gave him new housing and replaced his losses. In 1776 Euler's beloved wife Katharina died. A year later he married Katharina's half-sister Salome Abigail Gsell.

Almost immediately after he left Berlin, cataracts stole the vision from his left eye. An operation in 1771 briefly returned vision to this eye, but an infection caused a relapse and he became blind again. During this time Euler continued to publish mathematics, primarily by dictating his work to his son. Amazingly, Euler's mathematical output continued unabated. During this time of total blindness he proved some of his most important theorems and wrote some of his most influential books.

There is a widely held belief that a mathematician's most productive years are found in his youth; by the time he reaches forty—or thirty—all creativity and genius disappears. In his well-known memoir, *A Mathematician's Apology*, the British mathematician G. H. Hardy (1877–1947) wrote, "No mathematician should ever allow himself to forget that mathematics, more than any other art or science, is a young man's game."[15] While this does describe the declining quality of professional accomplishments for many mathematicians (and for practitioners of other creative fields), it does not represent the trajectory of Euler's career. His return to St. Petersburg was celebrated with fanfare, and he did not disappoint. As one historian wrote, Euler "demonstrated at once that he had not returned to Russia to retire but was, on the contrary, at the peak of his productivity."[16]

Just as Beethoven overcame the seemingly insurmountable obstacle of deafness to compose symphonies, so was Euler able to create deep, beautiful, and often "visual" mathematics from his darkened world. It is one of the great triumphs of the human spirit.

In addition to his pure mathematical research, Euler continued to make key contributions to applied mathematics. One of the most important problems of the day was to devise an accurate and reliable method of navigating at sea. Celestial navigation was only as good as the nautical tables that gave the locations of the heavenly bodies at any given time. The moon is the most conspicuous object in the nighttime sky, but since the motion of the moon is determined by the gravitational interaction of three bodies—itself, the earth, and the sun—it is an extremely difficult mathematical task to determine, in advance, its location at a specific time. Even today we do not fully understand the infamous three-body problem. Newton's theory of gravitation described planetary motion, but this work did not provide a computational algorithm for predicting this motion. In 1772 Euler developed a mathematical model of the motion of the moon that was computable and that yielded a very accurate approximation to the moon's motion. Extremely dependable sets of lunar tables were assembled from Euler's model. Expressing their thanks for his contribution, the Board of Longitude in France and the British parliament both rewarded Euler handsomely.

Euler's mathematical output continued until the day of his death, at the age of seventy-six. His last day was described by the Marquis de Condorcet (1743–1794) in his eulogy to Euler:

> He had retained all his facility of thought, and, apparently, all his mental vigour: no decay seemed to threaten the sciences with the sudden loss of their great ornament. On the 7th of September, 1783, after amusing himself with calculating on a slate the laws of the ascending motion of air-balloons, the recent discovery of which was then making a noise all over Europe, he dined with Mr. Lexell and his family, talked of Herschel's planet [the recently discovered planet Uranus], and of the calculations which determine its orbit. A little after he called his grand-child, and fell a playing with him as he drank tea, when suddenly, the pipe, which he held in his hand dropped from it, and he ceased to calculate and to breathe.[17]

Leonhard Euler is buried in St. Petersburg, Russia.

It is difficult to list Euler's greatest accomplishments in the subject of mathematics. We could quote one of his many theorems. We may also point to the successful textbooks that he penned, such as *Introductio in analysin infinitorum*, which the historian Carl Boyer called the most influential textbook in the modern era of mathematics. It might be his work in applied mathematics, such as his book *Mechanica* in which, for the first time, techniques of calculus are systematically applied to physics. It may be his writings for nonspecialists, such as the extremely popular *Letters to a German Princess*, which consists of a collection of lessons for Frederick the Great's niece, the Princess of Anhalt-Dessau. Perhaps it was his ability to organize and frame isolated results and seemingly unrelated ideas into a cohesive and ordered body of mathematics. Maybe it was the elegant and useful notation he created: Euler introduced e as the base of the natural logarithm; he made popular the use of the symbol π; at the end of his life he used i to denote $\sqrt{-1}$ (this notation was made popular by Gauss); he used a, b, and c to denote sides of a typical triangle with A, B, and C the angles opposite; he used Σ for sums; he denoted finite differences by Δx; and he began the use of $f(x)$ for a function.

It is difficult to single out one of Euler's many, many theorems as the most important. Some contend that it is the relation that brings 0, 1, π, e and i into one concise formula,

$$e^{\pi i} + 1 = 0.$$

Perhaps it is one of his wondrous infinite series formulas, which showed the power of calculus. It might be one of his theorems in number theory, such as those that brought closure to famous conjectures of Pierre de Fermat (1601–1665).

We, of course, will focus on the simple formula for polyhedra that relates the number of vertices, edges, and faces by

$$V - E + F = 2.$$

A recent survey of mathematicians showed that in their eyes, Euler's polyhedron formula is the second-most beautiful theorem in all of mathematics. The theorem voted most beautiful was Euler's formula $e^{\pi i} + 1 = 0$.[18]

In order to understand Euler's polyhedron formula we must look a little closer at polyhedra. What is a polyhedron?

CHAPTER 2

WHAT IS A POLYHEDRON?

*Madame, it is an old word and each one takes it new
and wears it out himself. It is a word that fills with meaning
as a bladder with air and the meaning goes out of it as quickly.
It may be punctured as a bladder is punctured and patched
and blown up again.*
—Ernest Hemingway, *Death in the Afternoon*[1]

According to the *Oxford English Dictionary*, the first appearance of the term "polyhedron" in English was in Sir Henry Billingsley's 1570 translation of Euclid's (c. 300 BCE) *Elements*. "Polyhedron" comes from the Greek roots *poly*, meaning many, and *hedra*, meaning seat. A polyhedron has many seats on which it can be set down. Although the term *hedra* originally meant seat, it has been the standard term for the face of a polyhedron since at least Archimedes.[2] Thus a reasonable translation of polyhedron is "many faces." By the time of Euler, the transliteration of *hedra* into Latin was well established.

Polyhedra are familiar three-dimensional geometric objects constructed from polygonal faces. Examples of polyhedra, shown in figure 2.1, include the common cube, the simple pyramid (known formally as the tetrahedron), the elegant icosahedron, and the soccer-ball-shaped truncated icosahedron.

Because of their beauty and symmetry, polyhedra have found a prominent place in art, architecture, jewelry, and games. Anyone who has walked by a store selling new-age items knows that some people believe polyhedra (crystals in particular) have mystical powers. Polyhedra appear in nature as well, taking the form of gemstones and single-celled organisms.

The properties of polyhedra have fascinated mathematicians for millennia. To prove theorems about polyhedra, it is crucial that we have a precise definition of the term. It was late in the development of the theory of polyhedra that anyone attempted to provide one. For many

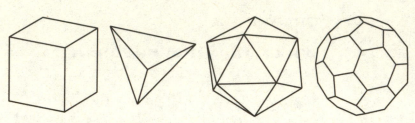

Figure 2.1. Examples of polyhedra.

years mathematicians operated under the "you know one when you see one" definition of polyhedron. They adopted the philosophy of Humpty Dumpty, who said to Alice, "When *I* use a word, it means just what I choose it to mean—nothing more nor less." That is never a good way to proceed. As Henri Poincaré (1854–1912) wrote:

> The objects occupying mathematicians were long ill defined; we thought we knew them because we represented them with the senses or the imagination; but we had of them only a rough image and not a precise concept upon which reasoning could take hold.[3]

Working without a proper definition can, and in this case did, generate theoretical inaccuracies and inconsistencies. As we will see, the rigor of Euler's proof of his polyhedron formula suffered because he did not explicitly define "polyhedron".

Agreeing on a suitable definition is surprisingly difficult. There have been many proposals over the centuries, not all of which are equivalent. Because of this inconsistency, there is no single definition of polyhedron that applies to the massive body of literature on these mathematical objects.

Naively we may define a polyhedron as a figure constructed from polygonal faces in which every edge is shared by exactly two faces, and from each vertex there emanates at least three edges. Indeed, this definition appears reasonable, but under close scrutiny we find that some such solids violate our intuitive sense of what a polyhedron should be. Whereas no one would disagree that the objects in figure 2.1 are polyhedra, we ask whether those in figure 2.2 (three of which satisfy our definition) should be so classified.

This is not a trick question. There is no historical consensus on whether the objects in figure 2.2 are polyhedra. The object on the far left, a cube with a corner removed, is a polyhedron according to most modern definitions; however the oldest definitions of polyhedron—those assumed implicitly

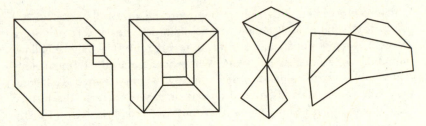

Figure 2.2. Shapes that are not convex polyhedra.

by the Greeks and Euler, for instance—do not allow a polyhedron to have indentations. Likewise, the second shape satisfies many mathematicians' criteria for being a polyhedron. But it has a tunnel running through it; it has the shape of a doughnut made from flat faces. Should we consider this a polyhedron? The third object consists of two polyhedra joined at a vertex, and the fourth is formed by joining two polyhedra along an edge. They are not, according to most definitions, polyhedra (although the third one does satisfy our criteria). Both figures have two interiors; if you were to fill them with water, you would have to fill two different compartments. And we could present other, even more pathological, examples that violate our sense of what a polyhedron should be.

For now we take the easy way out and sidestep this thorny task of giving a rigorous definition. Because we will present the historical development of Euler's formula, we can restrict our attention to a smaller class of polyhedra, one that is easier to define. We will take a very old-fashioned view of polyhedra, one with which Euler and the Greeks would agree. Although never stated explicitly, the historical assumption was that a polyhedron is convex. A *convex polyhedron* is an object that satisfies our naive definition (given above) and, in addition, has the property that any two points in the object can be joined by a straight line segment that is completely contained in the polyhedron. That is, a convex polyhedron cannot have any indentations. A quick inspection shows that every shape in figure 2.1 is convex and every shape in figure 2.2 is nonconvex.

We can see that this is what the Greeks assumed. They regarded the faces of a polyhedron as seats on which the polyhedron can rest. Every polyhedron of figure 2.1 can rest on any of its faces, but each polyhedron of figure 2.2 has at least one face that cannot be a seat of the polyhedron. Later, when we have more tools at our disposal, we will be able to apply Euler's formula to a broader class of polyhedra; but for the benefit of simplicity, and for historical reasons, we now consider only convex polyhedra.

Before proceeding, let's address one more historical discrepancy: is a polyhedron solid or is it hollow? Some definitions insist that polyhedra are solid, 3-dimensional objects, while others require that they be hollow, composed of a 2-dimensional "skin." One who assumes the former definition would construct a polyhedron from clay, whereas someone adopting the latter would make a polyhedron out of paper. Early in the history of polyhedra, the assumption was that they were solid. In fact, for many centuries polyhedra were called "solids." Later, as the theory of polyhedra made the transition to topology, the assumption of hollowness took hold. This assumption enabled theorems about polyhedra to be generalized to theorems about spheres and tori, which are, by definition, hollow. For most of what we have to say, either model will suffice. We will not make either assumption explicitly unless it is crucial to the discussion.

CHAPTER 3

THE FIVE PERFECT BODIES

There are always antecedent causes. A beginning is an artifice, and what recommends one over another is how much sense it makes of what follows.
—Ian McEwan, *Enduring Love*[1]

Modern geometry, and indeed much of modern mathematics, can trace its roots back to the work of the Greeks. During the period from Thales (c. 624–547 BCE) to the death of Apollonius (c. 262–190 BCE), the Greeks produced an astonishing body of mathematical works, and the names of many of the scholars of this era are familiar to schoolchildren everywhere: Pythagoras, Plato, Euclid, Archimedes, Zeno, and so on.

Although the mathematics from Egypt, Mesopotamia, China, and India may have influenced the Greeks, they quickly made the discipline their own. As Plato wrote in *Epinomis*, "Whenever Greeks borrow anything from non-Greeks, they finally carry it to a higher perfection."[2] Unlike earlier civilizations for whom utility was the prime objective, the Greeks wanted to understand the ideas of mathematics and have rigorous proofs of the assertions. Gone were the formulas used for approximations. Exactness, logic, and truth were the aims of their investigation.

The Greeks were fascinated with geometry, and their achievements in this field are far too numerous to list. It is certainly no exaggeration to say that most of the geometry now taught in schools was discovered by the Greeks. We shall focus on a Greek theorem about regular polyhedra (we will define "regular" shortly). It is one of the most celebrated and beautiful theorems in all of mathematics (fourth-most beautiful, according to the survey mentioned in chapter 1).

> There are exactly five regular polyhedra.

These five polyhedra are shown in figure 3.1. Three of the shapes are composed of equilateral triangular faces—the tetrahedron (the 4-sided

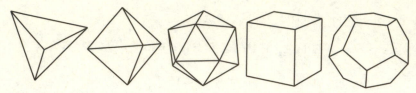

Figure 3.1. The five regular polyhedra: the tetrahedron, the octahedron, the icosahedron, the cube, and the dodecahedron.

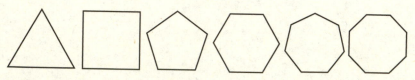

Figure 3.2. Regular polygons with 3, 4, 5, 6, 7, and 8 sides.

pyramid), the octahedron (the double-pyramid with 8 faces), and the 20-sided icosahedron. The cube is composed of 6 squares and the dodecahedron is the 12-sided figure made of regular pentagons. (Appendix A contains templates for making the regular polyhedra out of paper.)

The colorful history of these intriguing polyhedra begins with the Greeks and continues through the Renaissance to today. A proof that there are only five regular polyhedra is found in the final book of Euclid's *Elements*. (In chapter 8 we will present another proof using Euler's polyhedron formula.) Plato believed that the regular polyhedra were the building blocks of all matter. Because he incorporated them into an atomic theory, they are now called the Platonic solids. The astronomer Johannes Kepler (1571–1630) used the regular solids to create an early model of the solar system.

Beauty is often found in regularity, symmetry, and perfection. We are all familiar with the 2-dimensional *regular polygons*. A polygon is regular if every side has the same length and every interior angle has the same measure. The equilateral triangle is the only regular three-sided polygon, the square is the only regular four-sided polygon, and so forth (see figure 3.2). There are infinitely many regular polygons, one n-sided polygon for every integer n greater than 2.

The 3-dimensional analogue of a polygon is a polyhedron. The study of regularity in polyhedra is much more interesting than for polygons. While there are infinitely many regular polygons, the five figures shown in figure 3.1 are the only regular polyhedra.

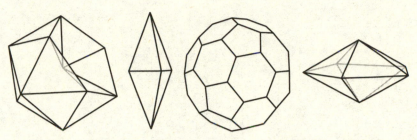

Figure 3.3. Polyhedra that are not regular. Each fails one of the four conditions for regularity.

What, exactly, are the criteria for a polyhedron to be regular? As with our definition of polyhedron, we must be careful not to include or exclude anything unintentionally. A *regular polyhedron* or *regular solid* is a polyhedron satisfying the following conditions.

1. The polyhedron is convex.
2. Every face is a regular polygon.
3. All of the faces are congruent (identical).
4. Every vertex is surrounded by the same number of faces.

Each of these criteria is necessary to obtain the desired collection of polyhedra. In figure 3.3 we give examples of polyhedra that fail to meet exactly one of the criteria. The first polyhedron satisfies every criterion except that it is nonconvex. The second polyhedron, a distorted octahedron, would be regular if the faces were equilateral triangles. The soccer ball is not regular because the faces are regular pentagons and regular hexagons. The last polyhedron consists of equilateral triangles, but four faces meet at all of the equatorial vertices, while five meet at the north and south poles.

Regular polyhedra are found in nature. Crystals are the most obvious naturally occurring polyhedra, and some crystals are regular. For instance, sodium chloride can take the form of a cube, sodium sulphantimoniate can take the form of a tetrahedron, and chrome alum can take the form of an octahedron. Pyrite, commonly known as fool's gold, can form a crystal with twelve pentagonal faces; however, the crystal is not a dodecahedron because the crystal structure creates pentagonal faces that are irregular.

In the 1880s, Ernst Haeckel of the HMS Challenger discovered and drew pictures of single-celled organisms called radiolaria. The skeletons of some of these organisms bear a striking resemblance to the regular polyhedra (figure 3.4).

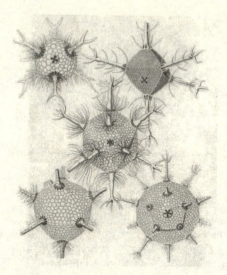

Figure 3.4. Radiolaria resembling the regular polyhedra.

There are some early man-made examples of regular solids. The cube and the tetrahedron, being relatively simple and common, have appeared in many physical creations throughout history. A dodecahedron dating back to at least 500 BCE was unearthed in an excavation on Mt. Loffa near Padua, Italy. An ancient icosahedral die was found in Egypt, but its origin is unknown.

What about the octahedron? This was probably the last of the five solids to be created by man. It is not as simple as the cube or the tetrahedron, so no everyday object would take this shape. It is not as exotic as the icosahedron or the dodecahedron, being nothing more than two square-based pyramids fused along their bases, so if it was encountered, it was likely ignored. Mathematics historian William Waterhouse argued that until someone discovered and put forth the notion of regularity, the octahedron was nothing special. He wrote, "The octahedron became an object of special mathematical study only when someone discovered a role for it to play."[3]

This discussion of the octahedron is enlightening. We see that there are three important stages in the development of the theory of regular polyhedra. The first is the construction of the objects themselves. Initially, their construction may be nothing more than making them out of clay, but eventually matters must become mathematical—they must be constructed geometrically. The second stage is the discovery of the abstract notion of regularity. This idea is obvious only in retrospect. Imagine handing the five regular solids to a passerby and asking, "what do these five objects have in common?" Chances are, he or she would not formulate the abstract notion of regularity. As Waterhouse said, "The discovery of this or that particular body was secondary; *the crucial discovery was the very concept of a regular solid.*"[4] Finally, the third stage is proving that there are only five regular solids. It must be shown, using rigorous mathematics, that there are five, and no more than five, of these beautiful objects. The development of this theory, from discovery, to abstraction, to proof, is due to the Greeks.

CHAPTER 4

THE PYTHAGOREAN BROTHERHOOD AND PLATO'S ATOMIC THEORY

[Pythagoras] was also the first to open up the enduring gulf of
incomprehension between the scientific spirit, which hopes that
the universe is ultimately understandable, and the mystical
spirit, which hopes—perhaps unconsciously—that it is not.
—George Simmons[1]

The early history of Greek mathematics is full of apocrypha, speculations, contradictory evidence, secondhand accounts, and just enough verifiable truths to create a fascinating puzzle. There are very few surviving records of Greek mathematics, and this scarcity of information makes it challenging to reconstruct the historical truth. Primary sources survived for several centuries after their creation, but nearly all were destroyed or lost during the Dark or Middle Ages. Much of what we know is not drawn from primary sources but from secondary sources written hundreds of years later.

We know little with certainty about Pythagoras (c. 560–480 BCE) and his band of followers, the Pythagoreans. As the philosopher W. Burkert wrote, "One is tempted to say that there is not a single detail in the life of Pythagoras that stands uncontradicted."[2] We believe the Pythagoreans were the first to study the regular solids. Pythagoras is said to have known of the cube and the tetrahedron, and there is a longstanding scholarly disagreement over whether he also knew of the icosahedron and the octahedron. One of his followers is credited with the discovery of the dodecahedron, and as we will see, the discovery may have resulted in his death.

Pythagoras was born on the Greek island of Samos in the Aegean Sea. According to some accounts, he traveled through Egypt and Babylon in his

Figure 4.1. Artist's rendering of Pythagoras.

early years and learned their mathematics and religion. Later he settled in the Greek city of Croton in what is now southern Italy.

Today Pythagoras is associated with the famous geometric theorem that bears his name,* but in his day he was known as a mystic and prophet. In Croton he became the spiritual leader of a secret society based on a philosophical religion. This was a time of religious significance in many cultures (he was a contemporary of Confucius, Buddha, and Lao Tzu). Pythagoras's successful brotherhood survived in Italy for nearly two hundred years after his death, and his doctrines continued to be taught until the sixth century CE. As time passed, Pythagoras's legend as a divine figure was enhanced by tales of miracles he performed.

In many ways the Pythagorean brotherhood was not unlike other cults of the day. The members were chosen very carefully—they participated in an initiation, endured ritual purifications, and took a vow of secrecy. They lived by a set of strict, sometimes bizarre, rules. According to legend, they were vegetarians, but were prohibited from eating beans; they could not stir a fire with a knife; they could not wear rings; and they had to touch the earth when it thundered.

The Pythagoreans believed in the transmigration of souls—that the souls of the dead returned as animals, and entered an infinite cycle of reincarnation moving up or down through the ranks of animals and

---

* The Pythagorean theorem states that if the legs and the hypotenuse of a right triangle have lengths a, b, and c, respectively, then $a^2 + b^2 = c^2$. In fact, this relationship was known to the Babylonians over a thousand years earlier.

humans. The only way to escape this cycle was through purification of body and mind. As in many cults, the purification of the body was achieved through modest living, abstinence, and restraint.

What distinguished the Pythagoreans was their means of purifying the mind. They did not achieve purity by meditating, but by studying mathematics and science. The ultimate union with the divine was said to follow from an understanding of the order of the universe, and the key to understanding the universe was to understand mathematics. Pythagoras said, "Beatitude is the knowledge of the perfection of the numbers of the soul."[3] This belief is expressed very succinctly by the Pythagoreans' motto, "All is number."

The Pythagoreans believed that God ordered the universe with numbers, and that every number could be expressed as a ratio of two whole numbers (every number could be written as a fraction). Using modern terminology, the Pythagoreans believed that all numbers are *rational*.

Music and astronomy were also important to the Pythagoreans. They discovered that musical intervals could be expressed as ratios and concluded that the most beautiful harmonies came from the most beautiful combinations of numbers. They contended that by using musical ratios one could explain astronomical phenomena such as the distances between the planets, the order of the planets, and the periods of revolution. They believed that the movement of the seven known planets (which included the earth, moon, and sun), like the vibrations of the seven strings of the lyre, created a harmony. Some claimed that this "music of the spheres" was audible to Pythagoras.

The Pythagoreans followed a communal lifestyle; they ate meals, exercised, and studied together. This way of living, along with their tradition of sharing knowledge orally, their enforced secrecy, and their adoration of Pythagoras, makes it difficult to discern which Pythagorean made what contribution to mathematics. In fact, since mathematics was part of their religion, and Pythagoras was their spiritual leader, any mathematical result obtained by his followers was "the word of the master" and was attributed to him.

According to legend, one of the Pythagoreans, Hippasus of Metapontum (c. 500 BCE), did not follow this tradition of deference, and was severely disciplined. One account states that he was drowned at sea, while another contends that he was expelled from the Pythagoreans and a tombstone was erected for him as a symbolic send-off. Again, accounts differ on what Hippasus had done to deserve this harsh treatment, and there are two competing stories (both of which may be true).

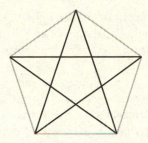

Figure 4.2. A pentagram, the symbol for the Pythagorean school, inscribed in a regular pentagon.

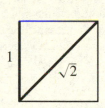

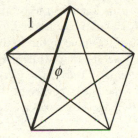

Figure 4.3. Diagonals of irrational length, $\sqrt{2}$ and $\phi = \frac{\sqrt{5}+1}{2}$.

One tale suggests that Hippasus discovered the dodecahedron and showed how to inscribe it in a sphere, but he failed to give credit to Pythagoras. This discovery may have been especially meaningful to the Pythagoreans because of the dodecahedron's pentagonal faces. They adopted the pentagram, or pentagon-star (see figure 4.2), which was the Greek symbol for health, as the special symbol used to identify others in the brotherhood. The pentagram is created by connecting the vertices of a regular pentagon, and doing so creates a new regular pentagon inside it.

The second tale states that Hippasus proved that not every number is rational, and he failed to keep this discovery a secret. Historians disagree over which irrational number Hippasus discovered. It could be $\sqrt{2}$, which is the length of the diagonal of a square with sides of length 1, or it could be $(\sqrt{5}+1)/2$, which is frequently called the *golden ratio*, or simply ϕ. That Hippasus discovered the irrationality of the golden ratio is an attractive theory because ϕ happens to be the length of the side of a pentagram that is inscribed in a pentagon with sides of length 1 (see figure 4.3). That all numbers are rational was one of the supporting pillars of the Pythagoreans' system of beliefs. The existence of an irrational number was a devastating and damaging realization. It is in this light that we can imagine their outrage at Hippasus. Ironically, the proof of the existence of an irrational number is one of the most significant and lasting contributions of the Pythagoreans to mathematics.

Regardless of whether it was Hippasus or one of his fellow Pythagoreans who discovered the dodecahedron, the Pythagoreans seemed to know of at least three of the regular solids: the tetrahedron, the cube, and the dodecahedron. It is not clear if they knew of the octahedron and the icosahedron, or if the discovery of these polyhedra should be attributed to Theaetetus of Athens (c. 417–369 BCE). Even early evidence is contradictory. Proclus (410–485), the fifth-century scholar, claims that the Pythagoreans knew of the octahedron and the icosahedron, whereas an undated scholium to Euclid's *Elements* states that "three of the aforesaid five figures [are] due to the Pythagoreans, namely the cube, the pyramid and the dodecahedron, while the octahedron and the icosahedron are due to Theaetetus."[4] Today, many scholars subscribe to William Waterhouse's theory of the late discovery of the octahedron, which would seem to rule out the Pythagoreans as discoverers.

Theaetetus is not as famous as some other Greek mathematicians, but he is unquestionably a hero of this story. He is almost certainly responsible for proving that there are five and only five regular polyhedra. Much of what we know about Theaetetus we know from the writings of his friend, the influential philosopher and teacher, Plato (427–347 BCE). Plato wrote two dialogues featuring Theaetetus: *The Sophist* and the eponymous *Theaetetus*.

Theaetetus was born during the Peloponnesian War. He died a war hero following the 369 BCE battle between Athens and Corinth. He studied mathematics under Theodorus (465–398 BCE), and by all accounts was a very gifted mathematician. Plato held Theaetetus in highest regard, second only to his teacher, Socrates (470–399 BCE). In Plato's *Theaetetus*, Theodorus says of the young Theaetetus, "This boy advances toward learning and investigation smoothly and surely and successfully, with perfect gentleness, like a stream of oil that flows without a sound, so that one marvels how he accomplishes all this at his age."[5]

At this time, the discovery of irrational numbers was a relatively recent event and not much was known about their properties. Theaetetus made important contributions to the classification and organization of irrationals. Later this classification would constitute most of Book X of Euclid's *Elements*.

Figure 4.4. Artist's rendering of Plato.

Despite the disagreement over who was the first to discover each of the five regular solids, there is little doubt that Theaetetus was the first to bring the study of the solids into a complete and rigorous form. Thanks to Theaetetus, the three phases of development that we discussed in chapter 3 were fulfilled. First, all five solids were known and Theatetus was able to construct them geometrically. Second, he perceived the common trait binding these five polyhedra—their regularity. Finally, he proved that these five solids are the only regular polyhedra. Theaetetus's proofs and constructions appeared later in Book XIII of Euclid's *Elements*. In fact, many historians contend that all of the mathematics in Books X and XIII of the *Elements* is due to Theaetetus.

Today Plato is best known as a philosopher and a writer, but one of his most important contributions was the creation of his school, the Academy. The Academy opened in the outskirts of Athens in approximately 288 BCE, a decade after the execution of Socrates. Plato founded the Academy to prepare young men for public life through the study of science and, especially, mathematics. It was his belief that by studying mathematics we learn to separate our intellect from our senses and our opinions. The Academy remained in existence for over nine hundred years. Its founding has been called "in some ways the most memorable event in the history of Western European science."[6]

Plato is not known for adding to the body of mathematics, but he was an important figure in the advancement of the subject. He was a lover of mathematics, and he held mathematicians in high regard. Mathematics formed the backbone of the curriculum at the Academy. This is clearly illustrated by the inscription found above its entrance: "Let no one ignorant of geometry enter here." Because many mathematicians were trained and nurtured in the Academy, it is said that Plato was not a maker of mathematics, but "a maker of mathematicians."[7]

As head of the Academy, Plato left much of the specific instruction to others. Theaetetus was one of these instructors, and it is believed that he taught at the Academy for fifteen years.[8]

It was from Theaetetus that Plato learned of the five regular solids. Plato recognized their mathematical importance and their beauty. Like many future thinkers, he believed that there must be cosmic significance for this magnificent collection of five objects. Plato was familiar with the view of the universe put forth by Empedocles (c. 492–432 BCE) stating that all matter is created from four primal elements: earth, air, fire, and water. These four building blocks feature prominently in Plato's *Timaeus*, a fictional account of a discussion between Socrates, Hermocrates, Critias, and Timaeus. Through a long monologue by the Pythagorean Timaeus of Locri, Plato laid out an elaborate atomic model, wherein each of the four elements, which Plato called bodies or corpuscles, is associated with one of the regular solids:

> To earth let us give the cube, because of the four kinds of bodies earth is the most immobile and the most pliable—which is what the solid whose faces are the most secure must of necessity turn out to be, more so than the others ... And of the solid figures that are left, we shall next assign the least mobile of them to water, to fire the most mobile, and to air the one in between. This means that the tiniest body belongs to fire, the largest to water, and the intermediate one to air—and also that the body with the sharpest edges belongs to fire, the next sharpest to air, and the third sharpest to water.

Using this rationale Timaeus concluded that fire is a tetrahedron, air an octahedron, and water an icosahedron. The fifth regular solid, the dodecahedron, could not be one of the elements. Timaeus contended that "this one the god used for the whole universe, embroidering figures on it."[9]

Timaeus went on to describe the interactions of the elements. The interactions are based on cutting and crushing; the sharper the element, the more it tends to cut, the less sharp, the more it tends to crush. This enables what we would describe as chemical reactions among fire, air, and water (but not earth, since it has square faces). The elements break apart and the triangular faces reform to create other elements. For example, one element of water (consisting of 20 equilateral triangles) can be broken apart and reformed to make three fire elements ($3 \cdot 4 = 12$ triangles) and one element of air (8 triangles). Timaeus noted that the different varieties of matter can be explained by the different sizes of the elements. He also addressed the phenomena of phase change: melting and freezing. For example, he would argue that metal is liquefiable water (as opposed to liquid water) that is composed of very large uniform icosahedrons which make it appear solid. When the sharp fiery tetrahedrons are applied to it, the icosahedrons separate from one another, the metal melts, and it is able to flow as a liquid.

The belief that earth, air, fire, and water are the four primal elements was adopted and expanded by Aristotle (384–322 BCE), a student of Plato. It was Aristotle who elevated the ether to be the fifth element, or *quintessence*, and argued that it was the material from which the heavenly bodies were made.

The Greek atomic model was so influential that it was universally accepted until the birth of modern chemistry two thousand years later. It was not until the Irish scientist Robert Boyle (1627–1691) published his book *The Sceptical Chymist* in 1661 that this chemical model began to crumble.

Although the Greek theory of chemistry is now a distant memory, the legacy remains. We still speak of being "exposed to the elements" when we step out into the wind (air) and rain (water). The elements appear implicitly and explicitly in many works of literature, pieces of artwork, mystical religions, fantasy games, and elsewhere. Some go so far as to argue that successful foursomes can be accurately paired with the elements (fire: John Lennon; water: Paul McCartney; air: George Harrison; earth: Ringo Starr).

Ever since Plato's treatment of the regular solids in *Timaeus*, the five regular polyhedra have been known as the *Platonic solids*.

CHAPTER 5

EUCLID AND HIS "ELEMENTS"

*At the age of eleven, I began Euclid, with my brother
as my tutor. This was one of the great events of my life, as
dazzling as first love. I had not imagined there was anything so
delicious in the world.*
—Bertrand Russell[1]

When one thinks of Greek geometry, one thinks of Euclid and of his masterwork, the *Elements*. In antiquity Euclid was often referred to simply as "the Geometer." It is disappointing that so little is known about his life. We cannot identify the place of his birth or even a reasonably accurate birth or death year. Most books on the history of mathematics do not venture a guess at his exact dates, saying instead that he was alive during the year 300 BCE.

Euclid learned mathematics and discovered the great works of Theaetetus and the other Platonists at Plato's Academy in Athens. Later in life he moved to Alexandria. This was during the time that the great library and museum were being constructed. Euclid founded a spectacularly successful and influential school of mathematics there.

Euclid wrote several books, but his eternal fame is due to one. In approximately 300 BCE he penned his magnum opus: the *Elements*. It was written as a textbook for elementary geometry, number theory, and geometric algebra. Euclid is not known for his new contributions to mathematics; much, if not most, of the material found in the *Elements* was first proved by others. Proclus wrote that Euclid "put together the elements, arranging in order many of Eudoxus' theorems, perfecting many of Theaetetus', and also bringing to irrefutable demonstration the things which had been only loosely proved by his predecessors."[2]

The *Elements* lacks much in the way of presentation; it does not place the mathematics in historical context, motivation is absent, and applications are not presented. However, the exposition and the logical treatment of the material were superior to anything that had come before it. Euclid began

Figure 5.1. Artist's rendering of Euclid.

with five seemingly "self-evident" assumptions, and based only on these simple postulates he developed the grand theories of geometry. Proclus praised the *Elements* as follows:

> [Euclid] included not everything which he could have said, but only such things as were suitable for the building up of the elements. He used all the various forms of deductive arguments, some getting their plausibility from first principles, some starting from demonstrations, but all irrefutable and accurate and in harmony with science ... Further, we must make mention of the continuity of the proofs, the disposition and arrangement of the things which precede and those which follow, and the power with which he treats each detail.[3]

This logical treatment fulfilled the dreams of Pythagoras from several centuries before. The impact to future scientists was profound. Armed with self-evident, fundamental truths, one tried to deduce all the laws of science. This ideal approach to science proved to be too simplistic; there are few laws of science akin to Euclid's five postulates. Nevertheless, the deductive, Euclidean approach to mathematics and science is still important today.

The *Elements* is the earliest major mathematical work created by the Greeks that has survived to this day. It was copied and recopied by hand numerous times until the first printed version appeared in Venice in 1482. Since then there have been an estimated one thousand printings.

Most of Book XIII, the final book of the *Elements,* is devoted to the Platonic solids. Some historians contend that the other twelve books were written only to prepare the reader for this final book. As we have said,

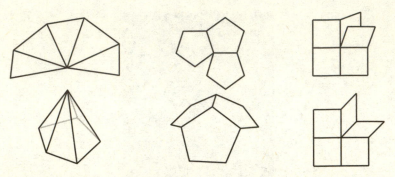

Figure 5.2. Splayed vertices from convex polyhedra (left and center), plus a non-convex polyhedron (right) for contrast.

the proofs found in Book XIII are most likely not due to Euclid but to Theaetetus. Some scholars contend that Euclid reprinted Theaetetus's work without editing it.[4]

The most important contribution of Book XIII is the proof that there are five and only five Platonic solids. First Euclid shows that there are at least five Platonic solids—that the tetrahedron, the octahedron, the icosahedron, the cube, and the dodecahedron are indeed regular. Then he proves that there are no more than these five. To accomplish the first of these tasks, Euclid gives explicit instructions on how to build each of the five Platonic solids—that is, he constructs the Platonic solids inside spheres. We will not repeat Euclid's constructions here. We will, however, present his argument that there are no more than these five. Later we give a different proof of this theorem, one that uses Euler's formula.

In his proof, Euclid uses a fact about plane angles. An angle in the face of a polyhedron is a *plane angle* (a cube has 24 plane angles measuring 90°). In Book XI Euclid proves that the plane angles that meet at any vertex of a convex polyhedron must sum to a value less than 360°. We omit the proof, but pictorially it is easy to see why the theorem is true. If we take the faces that meet at a vertex of a convex polyhedron and splay them out on a flat surface (to do so we must cut along one edge), the faces will not overlap one another and the two cut edges will not meet (see figure 5.2). This can only happen when the sum of the plane angles is strictly less than 360°.

Now consider a regular polyhedron. Each face is a regular polygon having n sides, and m edges of the polyhedron meet at each vertex. Because every face must have at least three sides, $n \geq 3$, and because at least three edges meet at each vertex, $m \geq 3$. Every angle of every face has

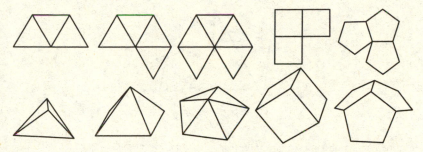

Figure 5.3. The five possible vertices for a Platonic solid, both splayed and unsplayed.

the same measure; call this angle θ. At each vertex there are m faces, each contributing a plane angle with measure θ. From Euclid's theorem, it follows that $m\theta$ must be less than $360°$. For which m and n is this possible?

When $n = 3$, the faces are equilateral triangles, so $\theta = 60°$. (The measure of an interior angle of a regular n-sided polygon is $180°(n-2)/n$.) Insisting that $m\theta < 360°$, we have $m(60°) < 360°$, or $m < 6$. So $m = 3, 4,$ or 5 are the only possibilities (see figure 5.3). These values of m yield the tetrahedron, the octahedron, and the icosahedron, respectively.

When $n = 4$, the faces are square, so $\theta = 90°$. This implies that $m(90°) < 360°$, or $m < 4$. So we can have only $m = 3$, and we obtain the cube.

When $n = 5$, the faces are regular pentagons and $\theta = 108°$. Thus $m(108°) < 360°$, or $m < 10/3$. So we can have only $m = 3$, and we obtain the dodecahedron.

When $n = 6$, the faces are regular hexagons and $\theta = 120°$. But $m(120°) < 360°$ implies $m < 3$, which is impossible. So there is no regular polyhedron with hexagonal faces. We encounter the same problem when $n > 6$. Thus there are no other Platonic solids.

Examining the proof, we see that Euclid overlooked some subtle details. In particular, he did not eliminate the possibility that there could exist two different polyhedra, both of which are made of regular n-gons and both of which have m faces meeting at each vertex. For instance, perhaps there is another polyhedron besides the icosahedron that is formed from equilateral triangles, five of which meet at each vertex. Euclid has the unstated assumption that this cannot happen. It turns out that Euclid is correct, so long as we are assuming convexity; but this fact needs to be proved. If we do not assume convexity, however, Euclid is wrong. In figure 5.4 we see a nonconvex polyhedron with the same properties as the icosahedron—it is composed of twenty equilateral triangles, five of which

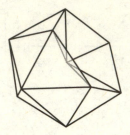

Figure 5.4. A nonconvex Platonic solid?

meet at each vertex. The only difference is that one of the vertices is pushed inward, making it nonconvex.

Pairs of polyhedra such as the icosahedron and the nonconvex icosahedron shown in figure 5.4 are called *stereoisomers* (borrowing a term from chemistry). They are constructed from identical collections of faces, and the faces are joined together along the same edges.

We must also consider the possibility that the polyhedra are flexible. Imagine making a polyhedron out of unbendable metal faces using hinges as edges. A conjecture that dates back at least to Euler is that such a polyhedron is not flexible even though all of the edges are hinged. Its shape cannot be changed by pulling, pushing, or squeezing. In 1766 Euler wrote that "[solid figures] can undergo change only to the extent that they are not undamaged or closed on all sides."[5] Proving this conjecture is important, because if one of the regular polyhedra is flexible, then we would have a whole family of stereoisomers, and thus an infinite number of subtly different regular polyhedra. This fact would destroy Euclid's proof.

It turns out that Euclid was correct, but the rigorous justification came two thousand years later from the prolific French mathematician Augustin-Louis Cauchy (1789–1857). In 1811, Cauchy proved that any two convex stereoisomers must be identical.[6] In other words, if we know the faces of a convex polyhedron and know which faces are neighbors, then we know the geometry of the polyhedron exactly. One consequence of this celebrated theorem is that the five Platonic solids are indeed unique. Another is that every convex hinged polyhedron is inflexible. This latter fact became known as the rigidity theorem for convex polyhedra. Remarkably, the rigidity conjecture does not hold for nonconvex hinged polyhedra, and this fact was not discovered until 1977. The American mathematician Robert Connelly constructed the first flexible nonconvex polyhedron.[7]

The Greeks' final major contribution to the theory of regular solids is due to Archimedes of Syracuse. Archimedes introduced the notion of

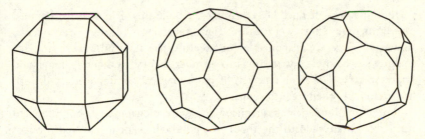

Figure 5.5. Three of the thirteen Archimedean solids.

semiregular solids. Like a regular solid, a semiregular solid is a convex polyhedron whose faces are regular polygons, but we now allow more than one type of regular polygon as a face. In addition, we insist that all faces with the same number of sides be congruent and that all of the vertices are identical (that is, each vertex has the same ordering of polygons surrounding it, and any vertex can be rotated to form any other vertex with the rest of the polyhedron lining up perfectly). Three semiregular polyhedra are shown in figure 5.5. Archimedes' work is lost, but according to the following passage by Pappus (c. 290–350 CE), we know that Archimedes found thirteen semiregular polyhedra:

> Although many solid figures having all kinds of faces can be conceived, those which appear to be regularly formed are most deserving of attention. These include not only the five figures found in the godlike Plato ... but also the solids, thirteen in number, which were discovered by Archimedes and are contained by equilateral and equiangular, but not similar, polygons.[8]

The complete set of thirteen polyhedra was reconstructed in 1619 by Kepler, who was unaware of Archimedes' work. Just as Theaetetus proved that the five Platonic solids are the only regular polyhedra, so did Kepler prove that there are only thirteen semiregular polyhedra. We should mention that there is an infinite collection of polyhedra called *prisms* and *antiprisms* that satisfy the semiregularity criteria, but historically these have not been called semiregular solids. Today the semiregular polyhedra are known as *Archimedean solids.*

Following the decline of Greek civilization, the center of mathematical activity moved to the Islamic empire. Under royal patronage, Arabic mathematicians translated many of the Greek mathematical classics, including works of Euclid, Archimedes, Apollonius, Diophantus, Pappus, and Ptolemy. They were more than caretakers of the Greek texts, however. They are responsible for creating the field of algebra and for making substantial contributions to number theory, number systems, and trigonometry. The Arabic period of mathematical dominance lasted until approximately the fifteenth century.

Arabic mathematicians advanced the state of geometry, but they did not substantially add to the theory of polyhedra. For a renewed interest in polyhedra, mathematics had to wait for Europe to emerge from the medieval period.

CHAPTER 6

KEPLER'S POLYHEDRAL UNIVERSE

Johannes Kepler is one of the great watershed figures in the history of science: half his mind churned with medieval fantasies and the other half was pregnant with the beginnings of the mathematicized science that formed the modern world.
—George Simmons[1]

During the period of Arabic mathematics, Europe was experiencing the darkness of the Middle Ages. Very few Europeans received a formal education; the great works of classical antiquity were all but forgotten; mathematical scholarship was almost nonexistent. Only the minimal teaching of geometry and arithmetic remained in monastic schools. For hundreds of years there were virtually no significant contributions made to the body of mathematics.

It was not until the European Renaissance in the fifteenth century that mathematical activity began to reemerge. The rise of the humanist movement brought a renewed interest in the Greek classics—first Greek literature, then Greek mathematics. The romance of Greek intellectual life is beautifully depicted in Raphael's fresco *School of Athens* (1510–1511), which features an imaginary gathering of Pythagoras, Euclid, Socrates, Aristotle, Plato, and other Greek scholars (figure 6.1).

Perspective was a prominent feature in Renaissance artwork. Polyhedra and the skeletons of polyhedra were excellent subjects for demonstrating an artist's mastery of perspective. Artists such as Piero della Francesca, Albrecht Dürer, and Daniele Barbaro contributed to both mathematics and art in their writings about perspective in polyhedra. Among the many artists who featured polyhedra in their artwork (see figures 6.2 and 6.3) are Leonardo da Vinci, who illustrated Fra Luca Pacioli's book *De divina proportione* (1509); Wentzel Jamnitzer, who created intricate and elaborate engravings of real and impossible polyhedra; Jacopo de Barbari,

Figure 6.1. Raphael's *School of Athens* (Fresco in the Apostotic Palace in the Vatican).

who painted a portrait of Luca Pacioli with his polyhedra; Paolo Uccello, who included polyhedra in his paintings and in the mosaics that he lay in the floor of the San Marco Basilica in Venice; Fra Giovanni da Verona, who created beautiful intarsia (wooden mosaics); and, as we will see (figures 6.5 to 6.8), Johannes Kepler, the physicist and mathematician.

Like the Renaissance scholars and artists of the two centuries preceding him, Kepler was fascinated by polyhedra. Today Kepler is most well known as an astronomer, famous for his three laws of planetary motion (describing the elliptical motion of the planets about the sun), but he made many other contributions to science and mathematics. He used the ideas of the infinite and the infinitesimally small in a way that anticipated calculus. He published work in the field of optics. He was an early user of logarithms. And Kepler made contributions, both real and fanciful to the theory of polyhedra.

Kepler was born December 27, 1571, in the small town of Weil der Stadt, Württemburg, in the Holy Roman Empire in present-day Germany. He had an extraordinarily difficult life: he was a sickly child raised in a troubled household, he endured religious persecution, his first wife and his favorite

Figure 6.2. Leonardo da Vinci's truncated icosahedron and pentakis dodecahedron from *De Divina Proportione*.

son died of smallpox, his mother was charged with witchcraft, and he died at the age of 58 while on a journey to recover his back salary. Despite these hardships, Kepler was a deeply religious man. He was headed for a career as a Lutheran minister until, at the age of twenty-three, he left the seminary to take a teaching position in mathematics and astronomy. His religious beliefs were important to him and, as can be seen in his writing, they often served as inspiration for his scientific work. As one of Kepler's biographers, Arthur Koestler, wrote, "This coexistence of the mystical and the empirical, of wild flights of thought and dogged, painstaking research, remained ... the main characteristic of Kepler from his early youth to his old age."[2]

Kepler believed that God created a world with mathematical beauty. Surely, Kepler believed, the existence of only five regular polyhedra must be significant; surely they must be reflected in the composition of the universe. Koestler wrote, "For Kepler's misguided belief in the five perfect bodies was not a passing fancy, but remained with him, in a modified version, to the end of his life, showing all the symptoms of a paranoid delusion; and yet it functioned as the *vigor motrix*, the spur of his immortal achievements."[3]

Figure 6.3. A marble inlay by Uccello (top left), one of Fra Giovanni's intarsia (top right), and samples from Wentzel Jamnitzer's *Perspectiva corporum regularium* (1568).

Figure 6.4. Johannes Kepler.

The inspiration for Kepler's first model of the solar system came on July 9, 1595, while he was lecturing to a room full of students. At this time the geocentric (Earth-centered) Ptolemaic view of the solar system was generally held to be the correct model. A half-century earlier Nicolaus Copernicus (1473–1543) had argued for the heliocentric (Sun-centered) model, but at this point, for various reasons, the heliocentric model was rejected by most intellectuals.

One day, as Kepler drew polygons inscribed in circles, he had the inspiration that this might be the secret to the planetary orbits: perhaps the orbits of the planets were nested circles inscribed in various polygons, with the sun at the center. After spending a summer meticulously working through the details, he realized that this was not the correct model for the solar system. Instead of abandoning this model entirely, he revised it and created one with which he was much happier. His new model appeared in his first book, *Mysterium Cosmographicum* (*Cosmic Mystery*) in 1596.[4]

Kepler had realized that polygons and circles were not the right objects for a model of the solar system; he shifted his attention up one dimension to polyhedra and spheres. He argued that the existence of the five Platonic solids must be related to the existence of the six known planets: Saturn, Jupiter, Mars, Earth, Venus, and Mercury. He claimed that the orbits of the planets are related to the nesting of the five Platonic solids inside spheres.

Take the sphere containing the orbit of the most distant planet, Saturn, along its equator. Inside this sphere inscribe a cube, and then, inside the cube, inscribe another sphere. Along the equator of this sphere, Kepler

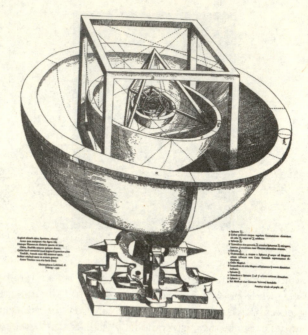

Figure 6.5. Kepler's early view of the solar system (from *Cosmic Mystery*).

contended, is the orbit of Jupiter (see figure 6.5). Continuing on in this manner (tetrahedron, sphere, dodecahedron, sphere, icosahedron, sphere, octahedron, sphere), we find the orbits of all six planets. Kepler wrote:

> This was the occasion and success of my labors. And how intense was my pleasure from this discovery can never be expressed in words. I no longer regretted the time wasted. Day and night I was consumed by the computing, to see whether this idea would agree with the Copernican orbits, or if my joy would be carried away by the wind. Within a few days everything worked, and I watched as one body after another fit precisely into its place among the planets.[5]

Thus, Kepler was the first professional astronomer to come out publicly, in print, in support of the Copernican model. At this time even Galileo (1564–1642), six years Kepler's senior, was silent in this matter.

The first half of *Cosmic Mystery* was very mystical—Kepler indulged in astrology, numerology, and symbology. He gave elaborate nonscientific reasons for why his model of the solar system was correct. He saw a very distinct hierarchy among the Platonic solids. For example, they were divided into the primaries (the tetrahedron, the cube, and the dodecahedron) and secondaries (the octahedron and the icosahedron), the primaries being those with three faces meeting at each vertex. He asserted that "containing is ...more perfect" than being contained,[6] and in his planetary model the primaries were the outer polyhedra and the secondaries the inner, with earth's orbit situated between the two classes.

Then, abruptly, in the second half of the book he switched to a scientific argument, complete with astronomical data. In order for his theory and the data to agree he made a few modifications to the model. Although he did not yet know that the orbits of the planets were ellipses, he knew that they were not circular. So, to contain the planets, each sphere in the model had to have some thickness; as a planet orbited in its noncircular pattern it remained inside its spherical shell. Kepler's model is surprisingly accurate; however, he realized that the data still did not quite fit the model (especially the orbits of Jupiter and Mercury). As a result, he found various ways to explain away these discrepancies, such as discrediting the data that he was using (which had come from Copernicus).

Later, Kepler himself proved that this archetype for the solar system was wrong. He wrote, "I indeed confess that the head of astronomy is struck off."[7] By sifting though a huge amount of data on Mars's orbit that were left to him by the astronomer Tycho Brahe (1546–1601), Kepler deduced the true motion of the planets. In one of the great feats in the history of science, he used these data to discover his three laws of planetary motion (the first two in 1609, the third in 1619). Thirty years after his death, these laws were mathematically verified by Isaac Newton. Despite the false claims in *Cosmic Mystery*, it is interesting that many of these outrageous ideas contain a grain of truth. Some of Kepler's greatest scientific accomplishments can be traced back to the seemingly nonsensical ideas in this book.

Kepler's most significant contribution to the theory of polyhedra came near the end of his career in his 1619 work *Harmonice mundi* (*The Harmony of the World.*)[8] This treatise has five parts; the first two are devoted to mathematics. In them Kepler discussed regular and semiregular polyhedra. He rediscovered all thirteen Archimedean solids and proved that there are no others. He presented the class of polyhedra called antiprisms. He also found a pair of star polyhedra, known today as the small and the great stellated dodecahedra (see figure 6.6). He called a polyhedron of this type an *echinus*, which means hedgehog or sea urchin. Later we will return

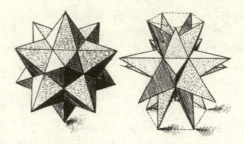

Figure 6.6. Kepler's drawings of the star polyhedra (from *The Harmony of the World*).

to these star polyhedra and see how they can be viewed as regular polyhedra and cause trouble for Euler's formula.

Even this late in his career, Kepler was fascinated by the Platonic solids. He subscribed to the Greek theory that there were four elements and to Plato's theory that they were made from the Platonic solids. We must remember that the publication of *The Harmony of the World* was still forty-two years before Boyle's revolutionary text, *The Sceptical Chymist*. In *The Harmony of the World* Kepler used ideas of Plato and Aristotle together with his own unscientific arguments to justify that the four elements were Platonic solids.

He argued that because a cube can be placed flat on a table in a way that it is not easily displaced, it is the most stable of the Platonic solids; thus it must be earth. An octahedron, held between two fingers, can be spun easily; thus it is the most unstable, and must be air. The tetrahedron contains the least volume for a fixed surface area; thus it is the driest of the five and must be fire. Similarly, the icosahedron contains the largest volume for a fixed surface area; so it is the wettest, and must be water. Kepler saw a relationship among the twelve faces of the dodecahedron and the twelve signs of the Greek Zodiac; because of this relationship, he argued that the dodecahedron must represent the universe. The correspondence between these elements and the Platonic solids can be seen in Kepler's famous illustration in figure 6.7.

In *The Harmony of the World,* we see again the dichotomy between Kepler's mystical tendencies and his brilliant scientific thought. In this work he made erroneous claims about atomic theory, but he also made an important new observation about the Platonic solids. He noticed that there is an antisymmetric relationship between the octahedron and the cube; an antisymmetric relationship between the dodecahedron and the icosahedron;

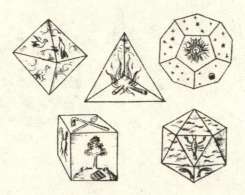

Figure 6.7. Kepler's drawings of the Platonic solids (from *The Harmony of the World*).

TABLE 6.1:
The number of vertices, edges, and faces in the Platonic solids.

	V	E	F
Octahedron	6	12	8
Cube	8	12	6
Icosahedron	12	30	20
Dodecahedron	20	30	12
Tetrahedron	4	6	4

and a self-symmetric relationship for the tetrahedron. As we can see in table 6.1, the cube and the octahedron both have 12 edges. The number of faces on the cube (6) is equal to the number of vertices on the octahedron, and the number of vertices on the cube (8) is equal to the number of faces on the octahedron. The same mirroring relationship holds for the icosahedron and the dodecahedron—both polyhedra have 30 edges, there are 20 faces on the icosahedron and 20 vertices on the dodecahedron, and there are 12 vertices on the icosahedron and 12 faces on the dodecahedron. The tetrahedron does not pair up with another regular polyhedron, but it has the same number of faces as vertices, so it pairs with itself.

Kepler recognized that this antisymmetry had a physical interpretation. Take one of the regular polyhedra, the cube, for instance. Place a new vertex at the center of each face. These eight points form the vertices of an octahedron. This new polyhedron is called the *dual* of the original polyhedron. In figure 6.8 we see Kepler's illustration that the octahedron

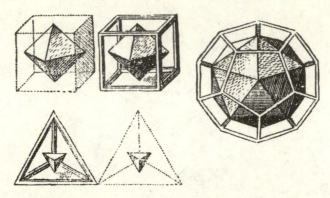

Figure 6.8. Kepler's depiction of dual polyhedra (from *The Harmony of The World*).

is the dual of the cube. Notice that each face of the cube corresponds to a vertex of the octahedron, so the number of faces of a cube is equal to the number of vertices of an octahedron. Examining the figures more closely, we see that each edge of the octahedron can be paired with an edge of the cube that is oriented at a right angle to it; thus both polyhedra must have the same number of edges. Moreover, each vertex of the cube corresponds to a face of the octahedron, and consequently there must be the same number of each. In this way we find the mirroring relationship present in table 6.1.

Similarly, Kepler showed that the icosahedron is the dual of the dodecahedron and that the tetrahedron is its own dual (see figure 6.8). Although Kepler knew that duality was reciprocal (the cube can be inscribed in the octahedron and the dodecahedron in the icosahedron), he did not show this. It did not fit with his hierarchy. Because he believed that containing was more perfect than being contained, he showed only primaries containing secondaries.

As was his style, Kepler could not resist sharing his unique interpretation of this mathematical observation. He assigned genders to the solids and used duality to indicate sexual compatibility. The cube and the dodecahedron (both dominant primaries) were male and contained the female octahedron and icosahedron (secondaries). The tetrahedron was a hermaphrodite because it contained itself. The faces and vertices were the sexual characteristics because that was where the solids met. Kepler wrote:

> However, there are two notable marriages, so to speak, of these
> figures, by combination from the two classes: the males, the cube and

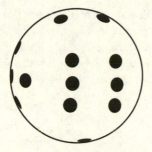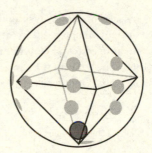

Figure 6.9. Round dice.

the dodecahedron from the primaries, the females, the octahedron and the icosahedron from the secondaries. In addition to these there is one which is, so to speak, celibate or hermaphrodite, the tetrahedron, because it is inscribed in itself, just as the feminine ones are inscribed in, and so to speak subject to, the males, and have the female tokens of their sex opposite to the masculine ones, or in other words the angles to the plane faces.[9]

Toy makers have creatively used the properties of regular and non-regular polyhedra to make many varieties of exotic dice. One ingenious toy maker has even used the duality of the regular polyhedra to make fair, usable round dice! The pips are painted on the surface of the sphere as if it was a cube (see figure 6.9). On the inside is a cavity in the shape of the cube's dual—an octahedron. A heavy ball bearing rattles around inside the octahedron until it comes to rest at one of its vertices. The weight of the ball bearing forces one of the die's "faces" to come to rest at the top of the die.

It is possible to extend the notion of duality to polyhedra that are not regular, although the definition is slightly more involved. Duality is a powerful, recurrent theme in mathematics. It is often the case that we create dual pairs by reversing some key quantity. For polyhedra, we are reversing dimension: zero-dimensional vertices replace 2-dimensional faces, and 2-dimensional faces replace zero-dimensional vertices. In other situations duality is obtained by reversing other quantities such as up and down, positive and negative, and so on. Sometimes the object that is most similar to a given object is the one that is its exact opposite. We will return to the notion of duality in chapter 23.

By the seventeenth century mathematics had become an active academic discipline in Europe. The long dry spell had ended. Polyhedra, reintroduced by the artist community, was again a subject of mathematical investigation. As we see in chapter 9, in approximately 1630 Descartes discovered important properties of polyhedra, but the world would not learn of this until 1860. The first major contribution to the theory of polyhedra in two thousand years would have to wait another century for Euler's brilliant insight.

CHAPTER 7

EULER'S GEM

"Obvious" is the most dangerous word in mathematics.
—E. T. Bell[1]

On November 14, 1750, the newspaper headlines should have read "Mathematician discovers edge of polyhedron!"

On that day Euler wrote from Berlin to his friend Christian Goldbach in St. Petersburg. In a phrase seemingly devoid of interesting mathematics, Euler described "the junctures where two faces come together along their sides, which, for lack of an accepted term, I call 'edges.'"[2] In reality, this empty-sounding definition was the first important stone laid in the foundation that would become a grand theory.

One of Euler's great gifts was his ability to consolidate isolated mathematical results and create a theoretical framework into which everything fits. In 1750 he set out to do this with polyhedra. He began what he hoped would be a study of the foundations of polyhedra, or *stereometry*, as he called it.

By then the theory of polyhedra was over two thousand years old, but it was purely geometric. Mathematicians focused exclusively on *metric* properties of polyhedra—properties that could be measured. They were interested in finding lengths of sides and diagonals; computing areas of faces; measuring plane angles; and determining volumes.

Euler's first step was not in this metric tradition. He hoped to discover a way to group together, or *classify*, all polyhedra by counting their features. After all, this is how we classify polygons—all three sided polygons are triangles, four-sided ones are quadrilaterals, and so on.

The difficulty of classifying polyhedra in this way is quickly apparent. The obvious feature to count—the number of faces—is not sufficient to distinguish a polyhedron. As we see in figure 7.1, polyhedra having the same number of faces can be quite different.

Euler's first brilliant insight was that the surface of a polyhedron is composed of 0-, 1-, and 2-dimensional components, namely vertices

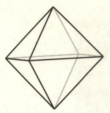 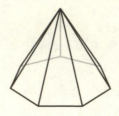 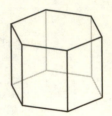

Figure 7.1. Three different eight-sided polyhedra.

(or solid angles, as he called them), edges, and faces, and that these were the features we must count. These three quantities became the standard building blocks of all topological surfaces. He wrote:

> Therefore three kinds of bounds are to be considered in any solid body; namely 1) points, 2) lines, and 3) surfaces, or, with the names specially used for this purpose: 1) solid angles, 2) edges, and 3) faces. These three kinds of bounds completely determine the solid.[3]

We cannot overstate the importance of this realization. Amazingly, until he gave them a name, no one had explicitly referred to the edges of a polyhedron. Euler, writing in Latin, used the word *acies* to mean edge. In "everyday Latin" *acies* is used for the sharp edge of a weapon, a beam of light, or an army lined up for battle. Giving a name to this obvious feature may seem to be a trivial point, but it is not. It was a crucial recognition that the 1-dimensional edge of a polyhedron is an essential concept.

For the faces of a polyhedron Euler used the well-established term *hedra*, which, as we have already mentioned, translates as "face" or "base." Euler referred to a vertex of a polyhedron as an *angulus solidus*, or solid angle. Before Euler wrote about polyhedra, solid angles were 3-dimensional entities defined by the faces that met at a point. A solid angle of a cube is different from a solid angle of a tetrahedron; they are distinguished by the geometry of the region they enclose. By the description given above—in which Euler associated a solid angle with a point—we see that he viewed the solid angles as zero-dimensional. When he said solid angle, he meant the very tip of the solid angle, not the 3-dimensional region that the faces enclose. The distinction is subtle, but the recognition that solid angles can be viewed as single points was crucial for his theorem. Nonetheless, Euler missed an opportunity by not giving them a new name. The vertex of a polyhedron *is* different from the solid angle on which it sits. In 1794

Figure 7.2. An East German stamp featuring Euler and his formula.

Adrien-Marie Legendre (1752–1833) made this point precisely:

> We often use the word *angle*, in common discourse, to designate the point situated at its vertex; this expression is faulty. It would be more clear and more exact to denote by a particular name, as that of *vertices*, the points situated at the vertices of the angles of a polygon, or of a polyhedron. In this sense is to be understood the expression *vertices of a polyedron* [sic], which we have used.[4]

Once the great Euler homed in on these three key features—vertices, edges, and faces—and started tallying their numbers for various families of polyhedra, he probably saw the relationship quickly. We can imagine Euler's surprise when he discovered that for any polyhedron, they satisfy the relationship

$$V - E + F = 2.$$

It is no wonder he expressed shock that no one had noticed it before. Brilliant Greek and Renaissance mathematicians had devoted countless hours to examining every conceivable aspect of polyhedra. How did they miss this elementary relationship?

The easy answer is the flippant retort that the history of mathematics is riddled with obvious theorems that went unnoticed for years. A more penetrating answer, however, is that the mathematicians who preceded him never viewed polyhedra in this way. Euler's predecessors were so focused on metric properties that they missed this fundamental interdependence. Not only did it not occur to them that they should count the features on a polyhedron, they did not even know which features to count.

Euler is indeed the master of us all.

Euler's work on the polyhedron formula was marked by three important documents. The first was his announcement to Goldbach in 1750 of his discovery of this relationship. He wrote:

> In every solid enclosed by plane faces the sum of the number of faces and the number of solid angles exceeds by two the number of edges, or $H + S = A + 2$.[5]

Euler used the letters H, A, and S to denote the number of faces (*hedra*), edges (*acies*), and vertices (*anguli solidi*). Renaming these quantities and rearranging the terms yields the familiar relationship:

> ### EULER'S POLYHEDRON FORMULA
> A polyhedron with V vertices, E edges, and F faces sastisfies $V - E + F = 2$.

In this letter Euler also included, without proof, ten other observations about polyhedra. He ended the letter by singling out the polyhedron formula and one other (which we will discuss in chapter 20) as the most important. Disappointed, he conceded that these two formulas "are so difficult that I have not yet been able to prove them in a satisfactory way."[6]

In 1750 and 1751 Euler wrote two papers about his polyhedron formula. Because of the slow turnaround of journal articles, they did not appear in print until 1758. In the first paper, "Elementa doctrinae solidorum" ("Elements of the doctrine of solids"),[7] he began his study of stereometry. For the first thirty pages, Euler made general remarks about polyhedra. Then Euler began his discussion of the relationship among the numbers of vertices, edges, and faces. He proved several theorems that relate V, E, and F and verified that $V - E + F = 2$ holds in several special cases. But he did not give a proof that the formula holds for all polyhedra. Still stymied and unable to complete the proof, he wrote, "I have not been able to find a firm proof of this theorem."[8]

The following year he published a second paper, "Demonstratio non-nullarum insignium proprietatum quibus solida hedris planis inclusa sunt praedita" ("Proof of some notable properties with which solids enclosed by plane faces are endowed").[9] Here he was finally able to give a proof of his polyhedron formula. Despite the fact that Euler's formula is one of the most famous in mathematics, his proof is virtually unknown by mathematicians today. There are several reasons for this. As we will see,

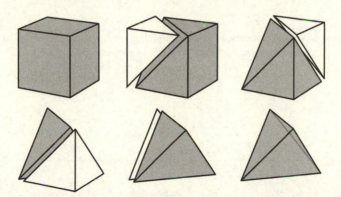

Figure 7.3. Removing vertices from a cube to obtain a tetrahedron.

Euler's demonstration does not satisfy the modern standards of rigor. Also, there have been many proofs of Euler's formula in the years since 1751 that are simpler and more transparent than Euler's. Still, Euler's proof is very clever and does not use metric properties of the polyhedra. The first truly rigorous justification was given four decades later, in 1794, by Legendre.[10] In his surprising argument, which we present in chapter 10, Legendre used geometric properties of spheres.

Euler's proof was a precursor to modern combinatorial proofs. He used the method of dissection to take a complicated polyhedron, perhaps containing many vertices, and systematically reduce it to a simpler polyhedron. Euler proposed that we remove vertices from the polyhedron, one at a time, until only four remain, leaving a triangular pyramid. By keeping track of the number of vertices, edges, and faces at each stage, and by using the known properties of the triangular pyramid, he was able to conclude that $V - E + F = 2$ for the original polyhedron.

Before jumping into Euler's proof, we look at an example. Consider the decomposition of the cube shown in figure 7.3. At each stage we remove a vertex of the cube by cutting away a triangular pyramid, continuing until we obtain a single triangular pyramid. Because the cube is a relatively simple polyhedron, we are able to remove each vertex by slicing away a single pyramid. In general, however, we may have to cut away several pyramids to remove a single vertex. Table 7.1 shows the number of vertices, edges, and faces at each stage of the decomposition.

One might hope that as the number of vertices decreases, the numbers of faces and edges decrease with some predictable pattern. As we can see in

TABLE 7.1:
The decomposition of a cube to a tetrahedron by removing vertices one at a time.

	Vertices	Edges	Faces	Edges — Faces
Cube	8	12	6	6
	7	12	7	5
	6	11	7	4
	5	9	6	3
Tetrahedron	4	6	4	2

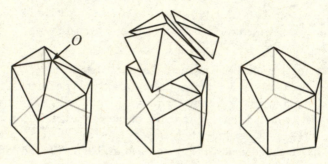

Figure 7.4. Remove the vertex O by cutting away pyramids.

the chart, however, the sequence has no obvious order. In this example the number of faces increases before decreasing—the polyhedron begins with six faces, then as the vertices are cut away the polyhedron has seven, then seven, then six, then four faces. This road seems to be a dead end. The key to Euler's proof is his astute observation that the difference between the number of edges and the number of faces decreases by one after a vertex is removed (as we see in the right-most column of the table). As we shall see, this is the heart of Euler's proof.

We begin with a polyhedron having V vertices, E edges, and F faces. Our first task is to remove a vertex from the polyhedron so that the resulting polyhedron has one fewer vertex than the original. After doing so, we must determine the number of faces and edges of the resulting polyhedron. Let O be the vertex that is to be removed, and suppose n faces (and therefore n edges) meet at O. Euler saw that O can be removed by cutting away $n-2$ triangular pyramids that have O as a vertex. For example, the polyhedron in figure 7.4 has a vertex formed from 5 faces, and it is removed by cutting away 3 pyramids.

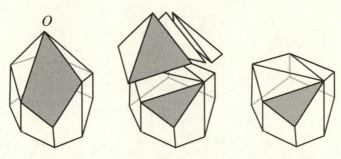

Figure 7.5. A nontriangular face contributes one new face and one new edge to the new polyhedron.

We would like to know the number of faces and the number of edges in the reduced polyhedron. As we saw with the cube, there is no simple answer. We must look at three special cases. We look at the simplest case first: assume that all of the faces meeting at O are triangular. By cutting away O we remove these n faces, but beneath the $n-2$ cut-away pyramids we find $n-2$ new triangular faces. Assuming that all of these new triangular faces lie in different planes, the number of faces in our new polyhedron is

$$F - n + (n-2) = F - 2,$$

where F is the original number of faces.

During this process we also remove the n edges that meet at the vertex O, but we add the $n-3$ edges that lie between the $n-2$ new triangular faces. Thus, the number of edges in the new polyhedron is

$$E - n + (n-3) = E - 3,$$

where E is the original number of edges.

Looking again at the example in figure 7.4, we began with a polyhedron having 11 faces and 20 edges. After removing the three pyramids we have a new polyhedron with $11 - 2 = 9$ faces and $20 - 3 = 17$ edges.

In the previous argument, we made two assumptions about the decomposition of the polyhedron. One was that all of the faces meeting at O are triangular, and the other was that the new triangular faces of the polyhedron are not coplanar. Now we must examine what happens if either or both of these assumptions do not hold.

Suppose one of the faces meeting O is not triangular (e.g. the shaded face in figure 7.5). Then, when the triangular pyramid that shares this face

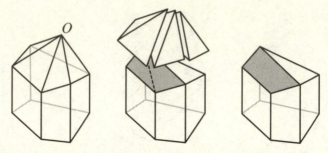

Figure 7.6. Two coplanar faces decrease the number of faces and edges by one.

is removed, the face does not completely disappear from the polyhedron. Also, a new edge is added where the face is cut in two. Thus, the number of faces and edges in the new polyhedron are both one larger than previously anticipated. In this example we begin with a polyhedron having 12 faces and 23 edges. After the three pyramids are removed, we obtain a polyhedron with $12 - 2 + 1 = 11$ faces and $23 - 3 + 1 = 21$ edges. In general, if the original polyhedron has s nontriangular faces meeting at O, then the number of faces and edges will be s larger than expected. So the number of faces is $F - 2 + s$ and the number of edges is $E - 3 + s$.

On the other hand, suppose that two of the new triangular faces are situated next to each other and lie in the same plane (e.g., the shaded faces in figure 7.6). Then they will not yield two distinct faces in the resulting polyhedron, but a single quadrilateral face. Thus, there will be one fewer face than was anticipated. Because there is no edge between these two faces, there will also be one fewer edge. In the example in figure 7.6 we begin with a polyhedron having 11 faces and 20 edges. After the pyramids are removed there are $11 - 2 - 1 = 8$ faces and $20 - 3 - 1 = 16$ edges. If this happens t times, then there will be t fewer faces and t fewer edges than anticipated. So, in the resulting polyhedron, the number of faces is $F - 2 + s - t$ and the number of edges is $E - 3 + s - t$.

These complicated-looking formulas represent the number of faces and the number of edges after a *single* vertex has been removed. The idea of keeping a running tally after several vertices are removed is daunting. However, Euler's important observation saves us from having to keep track of these values. If we take the number of edges on the new polyhedron and subtract the number of faces we obtain

$$(E - 3 + s - t) - (F - 2 + s - t) = E - F - 1.$$

In other words, the difference between the number of edges and the number of faces is exactly one less than it was before the vertex was removed. After n vertices have been removed, the difference in the number of edges and the number of faces is $E - F - n$.

With this, we can conclude Euler's proof. We began with a polyhedron with V vertices, E edges, and F faces. Suppose that we remove vertices one at a time, n in all, until only four vertices remain. Then $V - n = 4$, or $n = V - 4$. The only polyhedron with four vertices is a triangular pyramid (which has four faces and six edges). For a triangular pyramid, the difference in the number of edges and the number of faces is $6 - 4 = 2$, but from the previous discussion we know that it is also $E - F - n$. Thus we have the equations

$$E - F - n = 2$$

and

$$n = V - 4.$$

Substituting the second equation into the first and rearranging terms, we obtain $V - E + F = 2$, as desired.

At the outset, we claimed that Euler's proof is not completely rigorous and that he overlooked some subtleties. In fact, we see that Euler was very careful about keeping track of the numbers of faces and edges when a vertex was removed, However, he was too cavalier with the process of removing vertices, for he did not give detailed instructions about how to cut away the pyramids. Instead he made do with a few vague examples. Euler stated, correctly, that there may be several ways to remove a given vertex by cutting away pyramids, but he raised no warning flags that some decompositions are acceptable and others must be avoided. He left the reader with the incorrect impression that any decomposition will work as well as any other. In fact, some get us into trouble.

The first snag we encounter is that a decomposition may inadvertently yield a polyhedron that is nonconvex. Euler gave an example in which the vertex to be removed, O, has four adjacent vertices A, B, C, and D (see figure 7.7). He wrote:

> This can be done in two ways ... two pyramids will have to be cut away, either $OABC$ and $OACD$ or $OABD$ and $OBCD$. And if points A, B, C, D are not in the same plane the resulting solids will have a different shape accordingly.[11]

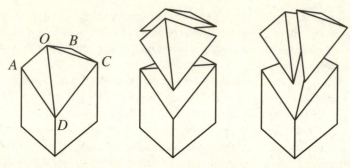

Figure 7.7. Removing a vertex from a polyhedron (left) may yield a convex (middle) or nonconvex (right) polyhedron.

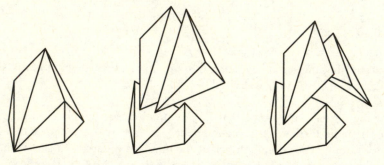

Figure 7.8. Euler's technique, applied to the polyhedron on the left may (middle) or may not (right) produce a degenerate polyhedron.

This is true, but if the four neighboring vertices are not coplanar, then one of the resulting polyhedra will necessarily be convex and the other one will be nonconvex. For the polyhedron in figure 7.7, removing pyramids $OABD$ and $OBCD$ will produce a nonconvex polyhedron.

Euler never mentioned convexity in his paper. He made the unstated assumption that all polyhedra are convex. If we look closely at his algorithm, we see that it is important that these polyhedra remain convex after a vertex has been cut away. Obtaining a nonconvex polyhedron can lead to trouble, for it may be impossible to use Euler's technique to remove a vertex located at a point of nonconvexity. Or we can encounter worse trouble than this.

As the mathematician Henri Lebesgue (1875–1941) pointed out, not only can the resulting polyhedron be nonconvex, it may not be a polyhedron at all![12] In figure 7.8 we see a vertex of a polyhedron that meets

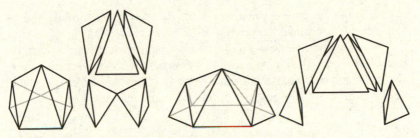

Figure 7.9. More problems with Euler's technique.

four faces. As with the previous example, we can remove this vertex in two different ways. One of the two methods works fine, but the other method yields a shape that is not a polyhedron, but rather an object consisting of two polyhedra joined along an edge. To make matters worse, this non-polyhedron does not satisfy Euler's formula ($V = 6$, $E = 11$, $F = 8$, so $V - E + F = 3$, not 2). This example seems to indicate a serious shortcoming in Euler's proof. In figure 7.9 we see that by applying Euler's dissections we can obtain other degenerate polyhedra. One decomposition yields two polyhedra joined at a vertex, and the second yields a disconnected polyhedron. Again, these objects fail to satisfy Euler's formula.

It turns out that Euler's proof is not beyond hope. With a little care we can repair his argument.[13] In all of the examples that we gave, making a wrong choice during the decomposition led to a breakdown of his proof; but in each case, there was a decomposition that was acceptable. We can prove that by making strategic, and not arbitrary, choices, we can remove a vertex in a way that *guarantees* that the resulting object is a convex polyhedron, thereby saving the day. So, after these repairs, we may finally assert that Euler's formula holds for all convex polyhedra.

Since Euler presented his proof, there have been many new proofs, most of which are more straightforward than his. We will see several of these in this book.

The subtle convexity problem turned out to be a real challenge for mathematicians to understand. It was the source of decades of interesting research, as mathematicians sought to determine exactly what properties a polyhedron must possess in order to satisfy Euler's formula. As we will see, they looked at nonconvex polyhedra, polyhedra with holes, and other, more pathological examples. This line of inquiry turned out to be extremely fruitful.

It took many years for mathematicians to see the importance of something that was apparent to Euler—that this theorem was about dimension and the rules for building mathematical objects. Euler's formula and its generalizations became the cornerstone for the field of topology.

It is likely that Euler had no idea of the importance of his theorem. He never returned to the problem of classifying polyhedra and he never wrote about the polyhedron formula again. He would never know that it would be one of his most beloved contributions to mathematics.

CHAPTER 8

PLATONIC SOLIDS, GOLF BALLS, FULLERENES, AND GEODESIC DOMES

Mathematics is concerned only with the enumeration and comparison of relations.
—Carl Friedrich Gauss[1]

Mathematicians do not study objects, but relations between objects.
—Henri Poincaré[2]

"That's great, but what's it good for?" the skeptical student asks, sarcasm dripping from his voice. Beauty is a wonderful trait, but some say usefulness is a more important measure of the worth of a theorem. What *is* Euler's formula good for?

That is a fair question to ask of any mathematical theorem. Euler's formula is more than just an elegant theorem. In the chapters that follow we will present many applications of Euler's formula. Most will require constructing the appropriate framework to understand the application. To whet the reader's appetite, we pause now and give two quick applications. First, we prove the uniqueness of the Platonic solids using Euler's formula, then we use Euler's formula to derive a structure theorem for golf balls, large molecules, and geodesic domes.

In chapter 5 we gave Euclid's proof that there are exactly five Platonic solids. Although his proof seems short, it relied on many of the geometric theorems proved in the previous twelve books of his *Elements*. In this chapter we give a different proof of the uniqueness of the five Platonic solids, one that uses only Euler's formula and a little arithmetic.

Suppose we have a regular solid. We will show that it must be one of the five known Platonic solids: the tetrahedron, the cube, the octahedron, the icosahedron, or the dodecahedron. Assume that the polyhedron has V vertices, E edges, and F faces. From Euler's formula we know that

$$V - E + F = 2.$$

Because the polyhedron is regular, each face is a regular polygon with the same number of edges. Obviously, this number, call it n, must be at least three. By definition the same number of edges meet at each vertex. This number, m, must also be at least three (of course, m is also the number of faces that meet at each vertex).

Each face contributes n edges, but because each edge is shared by two faces, the quantity Fn counts every edge twice. In other words,

$$E = \frac{1}{2}(Fn).$$

Similarly, each face contributes n vertices, but m faces meet at each vertex, so the quantity Fn counts each vertex m times. So,

$$V = \frac{Fn}{m}.$$

Now we substitute these quantities into the Euler's formula, and solve for F.

$$V - E + F = 2$$

$$\frac{Fn}{m} - \frac{Fn}{2} + F = 2$$

$$F\left(\frac{n}{m} - \frac{n}{2} + 1\right) = 2$$

$$F\left(\frac{2n - mn + 2m}{2m}\right) = 2$$

$$F = \frac{4m}{2n - mn + 2m}$$

We know that $4m$ and F are both positive. So for this last equation to be true, it must be the case that

$$2n - mn + 2m > 0.$$

It is easy to check that only five pairs of integers (n, m) satisfy this inequality, along with the requirements $n \geq 3$ and $m \geq 3$. Those are: $(3, 3)$,

TABLE 8.1:
Only five pairs of integers (n, m) satisfy the requirements for a regular polyhedron.

	n (edges per face)	m (faces per vertex)	$2n - mn + 2m$	V	E	F
Tetrahedron	3	3	3	4	6	4
Octahedron	3	4	2	6	12	8
Icosahedron	3	5	1	12	30	20
Cube	4	3	2	8	12	6
Dodecahedron	5	3	1	20	30	12

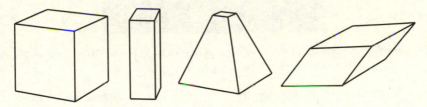

Figure 8.1. Cube-like shapes.

$(3, 4)$, $(3, 5)$, $(4, 3)$, and $(5, 3)$. Using the formulas for V, E, and F from above, we find that these correspond to the five Platonic solids (table 8.1).

We should reflect upon how surprising this proof is. Euclid's proof was local and geometrical. He used the angle measures of the regular faces to determine the possible configurations at the vertices. He used this local information to draw a conclusion about the global nature of the polyhedron.

This proof, on the other hand, is global and is virtually devoid of geometry. The theorem is about regular solids, but nowhere in the proof did we use the fact that the faces were regular polygons! We never even assumed that the faces were congruent. Euler's formula is combinatorial—it counts the numbers of vertices, edges, and faces. It is not possible to incorporate side lengths and angle measures into Euler's formula, yet we were able to use it to determine the Platonic solids.

Because we did not use all of the hypotheses of the theorem, we must have proved something different. What we did assume was that every face had the same number of sides, and that the same number of faces met at each vertex. From this point of view, all the shapes in figure 8.1 look the same—they all look like a cube.

Figure 8.2. A golf ball composed of 220 hexagons and 12 pentagons.

Essentially we proved that there are only five configurations of polyhedra with the property that every face has the same number of sides and the same number of faces meet at each vertex. Any such polyhedron must "resemble" a tetrahedron, an octahedron, an icosahedron, a cube, or a dodecahedron, just as the polyhedra in figure 8.1 all resemble a cube. In particular, the numbers of vertices, edges, and faces must be the same as one of the five Platonic solids.

In order to create golf balls that were more efficient flyers, one company invented polyhedral golf balls. Instead of a ball covered by round dimples, the exterior of their ball is constructed from 232 indented polygonal faces (see figure 8.2). A quick glance at the surface of the ball reveals a sea of hexagonal faces. Rest assured, however; this ball is not a sixth Platonic solid. Upon closer inspection, we find that 12 of the faces are pentagons.

In the introduction we learned about the family of ball-shaped, all-carbon molecules called fullerenes. In figure 8.3 we see the buckminsterfullerene, C_{60}, which has the same shape as a soccer ball. The carbon atoms form 12 pentagonal rings and 20 hexagonal ones. Scientists are able to create fullerenes with other numbers of carbon atoms. For example, C_{540} is a massive fullerene with 540 carbon atoms. The polyhedral structure of this molecule consists of 12 pentagons and 260 hexagons. In fact, every fullerene has pentagonal and hexagonal rings, and the number of pentagons is always 12.

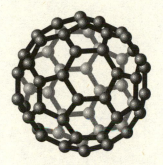

Figure 8.3. Fullerenes and soccer balls have exactly twelve pentagons.

The following theorem shows that this is not a coincidence. We use the term *degree* to indicate the number of edges meeting at a vertex.

TWELVE PENTAGON THEOREM

If every face of a polyhedron is a pentagon or a hexagon and if the degree of every vertex is three, then the polyhedron has exactly twelve pentagonal faces.

This theorem is a straightforward application of Euler's formula. Suppose we have such a polyhedron, with P pentagonal faces and H hexagonal ones. Since each pentagon has five sides and each hexagon has six sides, and since each edge borders two faces, the number of edges is $E = (5P + 6H)/2$. Likewise, since each vertex has degree 3, the number of vertices is $V = (5P + 6H)/3$. Inserting these quantities into Euler's formula yields

$$2 = V - E + F = (5P + 6H)/3 - (5P + 6H)/2 + (P + H).$$

By multiplying both sides by 6 we obtain the desired conclusion:

$$12 = 10P + 12H - 15P - 18H + 6P + 6H = P.$$

The twelve pentagon theorem has a dual formulation, which we obtain by interchanging the roles of the faces and vertices. We leave the proof of the theorem to the reader.

If every face of a polyhedron is a triangle and the degree of every vertex is five or six, then the polyhedron has exactly twelve vertices of degree five.

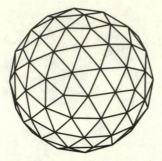 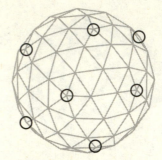

Figure 8.4. A geodesic dome with twelve vertices of degree five.

In figure 8.4 we see an example of such a polyhedron, with 7 of the 12 degree-five vertices highlighted. Many geodesic domes, such as the Biosphere in Montreal, are based on this design. Of course, architectural geodesic domes are not usually complete spheres. The Epcot Center in Disney World is based on this design, but each triangular face is subdivided into three more triangles.

With these simple examples we get our first glimpse of the power of Euler's formula. We see how a simple counting formula can force certain properties of polyhedra. In the chapters that follow we will see the great power behind Euler's seemingly elementary relation.

CHAPTER 9

SCOOPED BY DESCARTES?

*I hope that posterity will judge me kindly, not only
as to the things which I have explained, but also to those
which I have intentionally omitted so as to leave to others the
pleasure of discovery.*
—René Descartes[1]

In 1860, over a century after Euler presented his proof of the polyhedron formula, evidence surfaced that René Descartes, the famous philosopher, scientist, and mathematician, had known of this remarkable relationship in 1630, more than one hundred years before Euler. The evidence was found in a long-lost manuscript. The story is fascinating, as is the debate over whose name should accompany the polyhedron formula.

Descartes was born to a noble, if not wealthy, family in 1596 in La Haye, France, just outside of Tours. His mother died a few days after his birth, and his father, although supportive of his "little philosopher," was absent for much of René's childhood.

Young René was a sickly boy who developed into a hypochondriac as an adult. As a boy he attended the Jesuit school at La Flèche, and one of his teachers allowed him to remain in bed for as long as needed each morning, even when the other boys were attending lessons. Descartes used this time to think. He continued this practice throughout his life, nurturing many of his greatest ideas during the peaceful and quiet morning hours spent in bed.

A common theme in Descartes' life was his quest for solitude. As he put it, "I desire only tranquillity and repose."[2] This need for few distractions is reflected in his many relocations and his lifelong bachelorhood. During his stint in the army he enjoyed the extended periods of peace which afforded him quiet times for deep reflection. Although by no means a recluse, Descartes was always yearning for time alone to work on his scientific and philosophical pursuits. His motto illustrated this desire: *bene vixit qui bene latuit* (he has lived well who has hidden well).

Figure 9.1. René Descartes.

In 1637 Descartes published a short book with a long title, the influential *Discours de la méthode pour bien conduire sa raison et chercher la vérité dans les sciences* (*Discourse on the Method for Rightly Directing One's Reason and Searching for Truth in the Sciences*[3]). The publication of *Discourse on Method* marked the beginning of modern philosophy. In this book, which is now considered a literary classic, Descartes outlined a philosophy based on doubt and rationalism. It contains the most famous sentence in philosophy: *cogito ergo sum*, I think, therefore I am. His philosophy was one of the foundations of the Scientific Revolution.

Discourse on Method contained three appendices, the most important and influential of which, the hundred-page *La géométrie* (*The Geometry*), is frequently cited as the birth of analytic geometry, a subject so ingrained in today's mathematics that it is difficult to imagine working without it. (We should also give credit to Fermat, a contemporary of Descartes, for his contributions in this area.) Analytic geometry is the melding of geometry and algebra. In analytic geometry we introduce a coordinate system, and using it we locate a point by giving its coordinates (x, y). This system of coordinates is now called Cartesian coordinates, after Descartes. The power of this approach is that geometric figures—circles, lines, curves—can be represented by algebraic equations, enabling us to use the tools of algebra to solve geometric problems. Although the analytic geometry found in *The Geometry* is not fully formed (for instance, nowhere does Descartes explicitly create coordinate axes), many of the key ideas are present.

In 1649, after three years of repeated invitations, the fifty-three-year-old Descartes agreed to visit the young Swedish queen Christina and give her lessons in philosophy. The Queen insisted that the five-hour, thrice-weekly tutorials began at five o'clock in the morning and that they be held in an unusually cold room (during the coldest Swedish winter in sixty years).

The early morning lessons forced Descartes to abandon his long tradition of remaining in bed through the morning hours. These brutal conditions may have weakened Descartes' already frail body. On February 1, 1650, only months after he arrived in Sweden, he contracted pneumonia. He refused treatment from Christina's doctor, preferring to prescribe his own—a mixture of wine and tobacco that induced him to vomit the quickly gathering phlegm. His cure proved ineffective. He died on February 11, 1650.

Descartes' friend, the French ambassador Hector-Pierre Chanut, took it upon himself to ship Descartes' personal effects back to Paris, where they would be collected by Chanut's brother-in-law, Claude Clerselier. But the ship wrecked in the Seine, spilling its contents into the river. Descartes' possessions, including the trunk containing many pages of notes and manuscripts, floated away. Fortunately, three days later the trunk was recovered. The papers were carefully separated and hung to dry like the day's laundry.

Now in possession of Descartes' papers, Clerselier began publishing them. He also made the documents available for scholars to examine. Leibniz was one of the mathematicians interested in Descartes' waterlogged notes. On one of his trips to Paris, Leibniz made a copy of some of Descartes' notes about polyhedra, which dated back to roughly 1630. These important notes are now called *Progymnasmata de solidorum elementis* (*Exercises in the Elements of Solids*).

Clerselier died in 1684, eight years after Leibniz's visit, leaving manuscripts still unpublished. One of these was *The Elements of Solids*. The original was never seen again. Leibniz's personal copy disappeared and was not recovered for nearly two centuries. If not for providence, we would never know of Descartes' insightful work on polyhedra.

Foucher de Careil, a nineteenth-century Descartes scholar, was aware from Leibniz's letters that he had made copies of Descartes' missing manuscripts. In 1860 he searched for these documents in the well-organized Leibniz collection at the Royal Library of Hanover, but he did not find them. In a stroke of amazing good fortune, however, he found a dusty pile of unknown and uncataloged papers belonging to Leibniz in a neglected cupboard. It was in this collection that de Careil found Leibniz's copy of *The Elements of Solids*.

Like those who studied polyhedra before him, Descartes' approach was metric. Many of his formulas dealt with angle measures. Unlike his predecessors, but like Euler a century later, Descartes also took a combinatorial approach to polyhedra: he counted features on a polyhedron and created algebraic relations among them. While Euler would later tally the number of vertices, edges, and faces, finding the relationship $V - E + F = 2$, Descartes counted vertices (which, like Euler, he called solid angles), faces, and plane angles.

In his notes Descartes presented many facts about polyhedra. He gave no complete proofs, but it is not difficult to see how each formula follows logically from the ones before. The first major theorem generalized to polyhedra the well-known result for polygons that the sum of the exterior angles is 360°. We will discuss this result, now known as Descartes' formula, in detail in chapter 20. He also gave what may be the first algebraic proof that there are no more than five Platonic solids.

The work culminated in the following equality relating the number of faces, vertices, and plane angles (F, V, and P, respectively):

$$P = 2F + 2V - 4.$$

It is because of Descartes' discovery of this relation that some scholars say Euler's formula should bear his name. We simply observe that a polyhedron contains twice as many plane angles as edges (for example, a cube has 24 plane angles and 12 edges). That is, if there are E edges, then there are $P = 2E$ plane angles. Substituting $2E$ for P yields $2E = 2F + 2V - 4$. Dividing by two and rearranging terms, we obtain the familiar polyhedron formula.

So the questions arise: Did Descartes discover Euler's formula? If so, should it bear his name? The discovery of Descartes' notes sparked a debate that continues to the present day. Important mathematical personalities have disagreed on this topic. Even today, we find books that state emphatically that Descartes did or did not pre-discover Euler's formula. Of course, we should keep in mind the words of the eminent philosopher Thomas Kuhn (1922–1996), who wrote, "The fact that [the priority question] is asked . . . is a symptom of something askew in the image of science that gives discovery so fundamental a role."[4]

Ernest de Jonquières (1820–1901), one of the first and strongest supporters of Descartes, suggested that the theorem be called the Descartes–Euler formula. In 1890 he wrote, "It cannot be denied then that he knew it, since it is a deduction so direct and so simple, we say so intuitive, from the two theorems that he had just stated."[5] Jonquières' supporters argue that because Euler's formula follows so obviously from Descartes' work, either he knew the relationship, or he was close enough that the theorem should bear his name. They contend that if Descartes rewrote this sketch of a manuscript for publication, then he would have formulated the theorem in the now familiar way. Moreover, even if Descartes did not know the exact relationship, he proved a theorem that is logically equivalent to Euler's. He and Euler simply chose different quantities to relate. Today it is not uncommon for the polyhedron formula to be called the Descartes–Euler formula.

Surprisingly, much of the debate hinges on the concept of an edge of a polyhedron, which, as we have already discussed, was introduced by Euler. This feature is to us an obvious attribute, but it did not have a name in the time of Descartes. If one did look at the edge of a polyhedra, it was only as a side of one of the polygonal faces; the edges were the geometric objects used to create the angles. In order to give the usual form of Euler's formula, Descartes would have had to invent the notion of an edge.

Those who contend that Descartes did not anticipate Euler's formula maintain that the introduction of edges into the formula is crucial. As we saw earlier, Euler recognized that the essential importance of the theorem is that it relates zero-dimensional objects (vertices), 1-dimensional objects (edges), and 2-dimensional objects (faces). In the years that followed, Euler's formula was generalized to become an important theorem of topology. Topologists did not stop at 2-dimensional faces. As we will see in chapters 22 and 23, Poincaré and others extended Euler's formula to objects of any dimension.

All agree that Descartes was tantalizingly close, yet he failed to make the last important step. Plane angles are not the proper objects to compare to faces and vertices. To obtain the correct formulation he needed to introduce the notion of an edge. To those who say that Descartes must have known about the relationship with edges, critics point out that even the most accomplished mathematician can fail to see obvious consequences of his or her own work. After carefully examining the manuscript, the mathematician Henri Lebesgue wrote, "Descartes did not enunciate the theorem; he did not see it."[6]

There is a widely-held, mistaken belief that the objects in mathematics are named after their discoverers, and that when this does not happen, it is akin to plagiarism or falsification of history. Using this standard, Euler has been wronged repeatedly, for many of his discoveries bear the names of others. (There is an oft-repeated quip that "objects in mathematics are named after the first person after Euler to discover them.") There are countless examples (even in this book) of mathematical objects not named after the discoverer, but after someone who made influential contributions to the subject—perhaps the first person to truly recognize the importance of the discovery. Kuhn notes that, as in this instance, it is often the case that the priority of discovery is not clear-cut. "We so readily assume that discovering, like seeing or touching, should be unequivocally attributable to an individual and to a moment in time. But the latter attribution is always impossible, and the former often is as well... Discovering... involves recognizing both *that* something is and *what* it is."[7] (Recall Waterhouse's comment that the regular solids were not special until Theaetetus recognized the common trait that bound them together.)

Whether Descartes did or did not pre-discover Euler's formula is debatable. But because Descartes' work was never published and he did not recognize the "useful" form of the formula, it is not unreasonable that we continue to call $V - E + F = 2$ Euler's formula.

CHAPTER 10

LEGENDRE GETS IT RIGHT

*The bottom line for mathematicians is that the architecture
has to be right. In all the mathematics that I did, the essential
point was to find the right architecture. It's like building a
bridge. Once the main lines of the structure are right, then the
details miraculously fit. The problem is the overall design.*
—Freeman Dyson[1]

The second published proof of Euler's polyhedron formula, and the first
to meet today's rigorous standards, was given by Adrien-Marie Legendre.
Legendre was a French mathematician who belonged to both the Académie
des Sciences in Paris and the Royal Society of London. He published in
several areas, but his most important contributions were to number theory
and the theory of elliptic functions. His legacy also includes an extremely
popular textbook on elementary geometry that he wrote in 1794, *Éléments
de géométrie* (*Elements of Geometry*). In many ways Legendre's *Éléments*
replaced Euclid's *Elements*, becoming the primary geometry text for the
next hundred years, and it became the prototype for future generations
of geometry textbooks. It was translated into English several times; one
American translation had thirty-three editions.

Legendre included Euler's polyhedron formula in *Elements of Geometry*, and the book's popularity gave it wide exposure. Legendre did not
repair Euler's proof; instead, he presented a new demonstration—one that
differed substantially from Euler's. Legendre's ingenious argument used
concepts from spherical geometry and metric properties such as angle
measures and area. The success of such a proof is especially unexpected
because these concepts do not appear in the statement of the theorem.

The key to Legendre's proof is an elegant formula from spherical
geometry that gives the area of a triangle on the surface of a sphere in
terms of its interior angles. On a sphere, triangles and other polygonal
figures are not made from straight lines but from arcs of great circles.
A *great circle* is any circle on a sphere that has the same radius as the

Figure 10.1. Adrien-Marie Legendre.

Figure 10.2. Lay a ribbon on a sphere to find a great circle.

sphere, or, equivalently, it is any circle on the sphere of maximum possible radius. Examples of great circles on the Earth are the equator and lines of longitude. Lines of latitude other than the equator—the Tropics of Cancer and Capricorn, the Arctic Circle—are not great circles. Great circles are not straight, but they are as close to straight as is possible on a sphere. They have the noteworthy property that they are length-minimizing. That is, the shortest path between two points on a sphere follows a great circle. Ignoring physical realities such as wind currents and the rotation of the earth, an airplane flying from Pennsylvania to India seeking the shortest route would fly along a great circle that passes over Iceland.

A practical way to find great circles on a small sphere is to use a ribbon (figure 10.2). Take a wide ribbon such as those on birthday presents and lay it on the sphere. Wrap the ribbon around in such a way that it lies flat and does not twist laterally. When we do this the ribbon traces out a great circle.

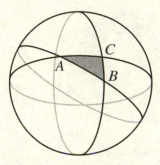

Figure 10.3. A triangle formed by three great circles.

We define a triangle on a sphere to be a region bounded by three great circles (as shown in figure 10.3). In mathematical language a great circle is known as a *geodesic*, so a more precise name for a spherical triangle is a *geodesic triangle*. We insist, as Legendre did, that each side of a geodesic triangle is less than half the circumference of the sphere.

Geodesic triangles were first introduced by the Greek mathematician Menelaus of Alexandria (c. 98) in his book *Sphaerica*. In this book Menelaus presented a theory for spherical geometry that is analogous to Euclid's theory for planar geometry found in *Elements*. He showed that many of the theorems for planar triangles are also true for geodesic triangles. For instance, the sum of the lengths of two sides of a spherical triangle is always greater than the length of the third side. He proved an interesting result that is true on a sphere, but not in a plane: two similar geodesic triangles (they have angles of equal measure) must be congruent. On the other hand, one of the most well-known theorems from planar geometry—that the sum of the interior angles of a triangle is 180°, or π radians—does not hold on a sphere.* On a sphere, the sum of the interior angles is always greater than π. For example, the large geodesic triangle shown in figure 10.4 has three right angles; so they sum to $3\pi/2$. The smaller geodesic triangle experiences less of the curvature of the sphere, so the angle sum is smaller—but it still exceeds π.

For nearly fifteen hundred years no one refined Menelaus's statement about the sum of the interior angles. It was not until the seventeenth

---

*In everyday life we measure angles in degrees—a right angle is 90°, there are 360° in a circle, and so on. However, in most mathematical applications angles are measured in radians. The conversion is simple: 180° corresponds to π radians. So a right angle is $\pi/2$ radians and there are 2π radians in a full turn. We will see concrete examples later that illustrate why radians are superior to degrees.

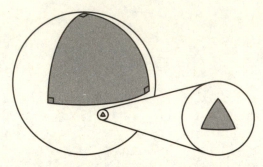

Figure 10.4. Geodesic triangles on a sphere.

century that two men, Thomas Harriot (c. 1560–1621) and Albert Girard (1595–1632), quantified the excess in the angle sum.

In figure 10.4 we see that there is a direct relationship between the area of a triangle and the sum of its interior angles. As a triangle grows in size, it becomes more distorted because of the curvature, and the sum of the interior angles grows.

The theorem of Harriot and Girard gives a formula relating three quantities: the sum of the interior angles of a geodesic triangle, the area of the triangle, and the radius of the sphere on which it lies. For simplicity, we shall give the formula for triangles on the *unit sphere*—the sphere of radius one. (The formula for a sphere of a different radius is obtained by scaling the quantities appropriately.)

HARRIOT-GIRARD THEOREM

The area of a geodesic triangle on the unit sphere with interior angles a, b, and c is $a + b + c - \pi$. In other words, area = (angle sum) $- \pi$.

Because the sum of the interior angles of every planar triangle is π, we can restate the formula in yet another way:

$$\text{area} = (\text{angle sum}) - (\text{angle sum for a planar triangle}).$$

That is, the area of a spherical triangle is precisely the amount that the angle sum exceeds the angle sum of a planar triangle. As we will see, this remarkable formula generalizes to spherical polygons with more than three sides. By the way, this is our first concrete example showing why it is advantageous to measure angles in radians; it is invalid when angles are measured in degrees.

As a warm up exercise, let us verify that this theorem holds for the large geodesic triangle shown in figure 10.4 (assuming that the sphere is the unit sphere). We can cover the sphere with eight of these triangles—four in the northern hemisphere and four more in the southern hemisphere. So the area of this triangle is one-eighth the area of the sphere. Because the surface area of a sphere of radius r is $4\pi r^2$, the unit sphere ($r = 1$) has surface area 4π. Hence, the area of the triangle is one-eighth of 4π, or $\pi/2$.

Now we can check that the Harriot-Girard theorem yields the same value. The sum of the three interior angles of this triangle is $3\pi/2$. So, according to the theorem, the area of this triangle must be $(3\pi/2) - \pi = \pi/2$, which agrees with our previous calculation.

This relationship was discovered independently by Harriot and Girard. The British scholar Thomas Harriot is somewhat of an enigma. He was a talented and active researcher, but he never published any of his work. When he died, he left ten thousand pages of unpublished manuscripts, diagrams, collections of measurements, and calculations. One biographer wrote that Harriot's aversion to publishing "may largely be explained by adverse external circumstances, procrastination, and his reluctance to publish a tract when he thought that further work might improve it."[2] Many of his papers appeared posthumously. He is best known for his work in algebra, but he also studied optics, astronomy, chemistry, and linguistics. Harriot, like Leibniz and Euler, had a reputation for introducing new and elegant mathematical notation. Unfortunately, because of the difficulty of typesetting his nonstandard notation, not all of it appeared in print and was thus not widely adopted. Two symbols that did survive are the signs $<$ for "less than" and $>$ for "greater than." Very little is known about Harriot's personal life. In 1585 he was sent by Sir Walter Raleigh on a one-year trip to the New World as a surveyor and mapmaker. So it is likely that he was the first established mathematician to set foot in North America.

The French mathematician Albert Girard resided in Holland, most likely because he was uncomfortable living in his childhood home of Lorraine, France, as a Protestant. Today he is known for his work in algebra and trigonometry. He has the distinction of being the first to use the abbreviations sin, tan, and sec for the trigonometric functions sine, tangent, and secant, and for introducing the notation $\sqrt[3]{\ }$ for cube root. Girard was also the first mathematician to assign a geometric significance to negative numbers. He wrote, "The negative solution is explained in geometry by moving backward, and the minus sign moves back when the + advances."[3]

Historically, the area formula for spherical triangles had Girard's name attached to it, and not Harriot's. This attribution is understandable because

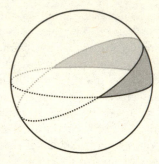

Figure 10.5. A lune on a sphere.

Girard's proof, which was published in 1629,[4] was the first to appear in print. Girard was known for his terse writing style, and his proofs often suffered from lack of detail. This proof was even unsatisfactory to Girard, himself—he described the result as "a probable conclusion."[5] Unbeknownst to him, the same theorem had been proved by Harriot twenty-six years earlier. Of course, as we have noted, Harriot did not publish this result, or any mathematical result. Nonetheless, he did not keep it a secret. His proof was known to contemporaries; the British mathematician Henry Briggs (1561–1630) had told Kepler of Harriot's result and had included it in a list of the great discoveries of his time. However, there is no indication that Girard was aware of Harriot's proof.

Because Harriot was the first to prove the theorem and Girard was the first to publish it, we now call the result the Harriot-Girard theorem. It is worth noting that Harriot's proof was much simpler and more elegant than was Girard's. The argument we give below is due to Legendre, but it is very similar to Harriot's.

Legendre's proof cleverly uses an object called a *lune* (occasionally called a *biangle* in analogy with a triangle). A lune is a region bounded by two great circles (see figure 10.5). A pair of great circles always meet at two points that lie on the exact opposite sides of the sphere. If the two circles meet with an angle a at one end, then they also meet with angle a at the other end. If the angle a is measured in radians, then the area of the lune (on a unit sphere) is $2a$. We easily justify this fact by setting up a simple ratio: the area of the lune is to the total surface area of the sphere as a is to 2π (as can be seen in figure 10.6). So, we have

$$\frac{\text{Area of lune}}{4\pi} = \frac{a}{2\pi}.$$

Solving the equation for the area we find it to be $2a$.

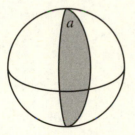
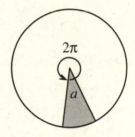

Figure 10.6. A lune on a sphere (left) and the view from above (right).

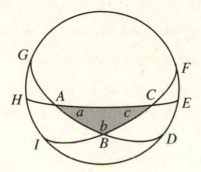

Figure 10.7. Great circles in a hemisphere.

Now, consider a geodesic triangle ABC on a unit sphere with interior angles a, b, and c. The triangle is contained in some hemisphere. Extend the sides of ABC to meet the boundary of this hemisphere. As in figure 10.7, let D, E, F, G, H, and I denote the points where these lines meet the hemisphere's edge.

By the symmetries of the sphere, the sum of the areas of the regions ADE and AGH is the same as the area of a lune with angle a. In other words, if we were to cut out the triangle AGH and glue edge GH to edge ED, then we would have a lune with angle a. From this observation we conclude that

$$\text{area}(ADE) + \text{area}(AGH) = \text{area of lune} = 2a.$$

Similarly, triangles BFG and BDI have the same area as a lune with angle b, and triangles CHI and CEF have the same area as a lune with angle c. So we have

$$\text{area}(BFG) + \text{area}(BDI) = 2b$$

and

$$\text{area}(CHI) + \text{area}(CEF) = 2c.$$

Summing these three equations gives

$$[\text{area}(ADE) + \text{area}(AGH)] + [\text{area}(BFG) + \text{area}(BDI)] +$$
$$[\text{area}(CHI) + \text{area}(CEF)] = 2a + 2b + 2c.$$

Looking closely at the left-hand side of this expression, we see that we are adding the area of every region of the hemisphere once, except the triangular region ABC, which we are adding three times. So we have

$$\text{area}(\text{hemisphere}) + 2 \cdot \text{area}(ABC) = 2a + 2b + 2c.$$

Because the area of the hemisphere is 2π, we obtain

$$2\pi + 2 \cdot \text{area}(ABC) = 2a + 2b + 2c.$$

Rearranging this expression and dividing by 2, we conclude that

$$\text{area}(ABC) = a + b + c - \pi,$$

as desired.

For Legendre's proof of Euler's formula we need the following generalization of the Harriot-Girard theorem to geodesic polygons with more sides than three.

HARRIOT-GIRARD THEOREM FOR GEODESIC POLYGONS

The area of an n-sided geodesic polygon on the unit sphere with interior angles $a_1, a_2, \ldots, a_n$ is $a_1 + \cdots + a_n - n\pi + 2\pi$, or equivalently, area = (angle sum) $- n\pi + 2\pi$.

The sum of the interior angles of any planar polygon with n sides is $(n-2)\pi$. (We will look closely at this theorem and its generalizations in chapter 20.) So, just as with triangles, the area of a geodesic polygon is simply the amount that its angle sum exceeds the angle sum of a planar polygon with the same number of sides. That is,

$$\text{area} = (\text{angle sum}) - (\text{angle sum for a planar } n\text{-gon}).$$

To see why this generalization holds, divide the polygon into geodesic triangles by adding diagonals. Such a decomposition has $n - 2$ triangles

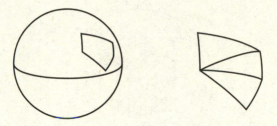

Figure 10.8. A polygon on a sphere divided into triangles.

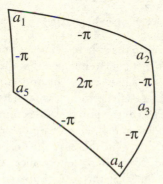

Figure 10.9. The area of the spherical polygon is the sum of the quantities on the diagram.

(see figure 10.8). The sum of the areas of these triangles is the area of the polygon, and the sum of the angles of the triangles equals the sum of the angles of the polygon. Applying the Harriot-Girard theorem to all $n-2$ triangles and summing, we find the area of the polygon to be

$$\text{area} = a_1 + \cdots + a_n - (n-2)\pi = a_1 + \cdots + a_n - n\pi + 2\pi.$$

There is an easy way to remember this formula. Visualize the polygon as shown in figure 10.9. Place the angle measure at each angle, $-\pi$ on each edge, and 2π in the middle of the face. The area of the polygon is simply the sum of these quantities. This visual representation will be useful in understanding Legendre's proof of Euler's formula.

Finally, we are ready to give Legendre's argument. Begin with a convex polyhedron with V vertices, E edges, and F faces. Let x be any point in the interior. As shown in figure 10.10, construct a sphere centered at x that surrounds the polyhedron completely. Because the units are unimportant, we may choose them so that the sphere has radius one.

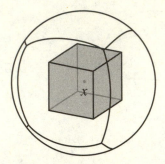

Figure 10.10. The projection of a polyhedron onto a sphere.

Project the polyhedron onto the sphere using rays emanating from x. One way to visualize the projection is to imagine that the polyhedron is a wire-framed model and that x is a light bulb. The projection is the shadow of the wire frame on the surface of the encompassing sphere. We will not prove it, but in this case the faces of the polyhedron become geodesic polygons.

In his proof, Legendre employed a common mathematical trick. He computed the same quantity—in this case the area of the unit sphere—in two different ways, thereby deriving an equality. First, he used the well-known area formula to find that the surface area of the unit sphere is 4π. Second, he summed the areas of each face on the sphere which, clearly, also gives the overall surface area.

By the Harriot-Girard theorem we know that the area of each n-sided face is the sum of the interior angles minus $n\pi - 2\pi$. Rather than working with this formula directly, we use the visual representation introduced in figure 10.9. Label all of the angles, edges, and faces on the sphere—placing the angle measure at each angle, $-\pi$ on both sides of every edge, and 2π in the middle of every face—yielding a sphere such as the one in figure 10.11. To compute the surface area of the sphere we sum all of the labeled quantities.

Although the angles meeting at a vertex of the polyhedron sum to less than 2π, these same angles sum to 2π when projected onto the smooth surface of the sphere. Since there are V vertices, the vertices contribute $2\pi V$ to the sum. Each edge contributes -2π; namely, $-\pi$ from one side and $-\pi$ from the other side. Since there are E edges, they contribute $-2\pi E$ to the sum. The middle of each face is labeled with 2π. Since there are F faces, they contribute $2\pi F$ to the sum. Putting all of this together we find that the total surface area of the sphere is

$$4\pi = 2\pi V - 2\pi E + 2\pi F.$$

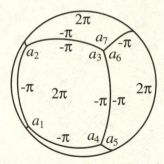

Figure 10.11. The projection with labels.

Dividing by 2π, we obtain Euler's formula,

$$2 = V - E + F.$$

Even a quick comparison of Euler and Legendre is enough to show that their proofs are very different. On the one hand, Euler's seems to be "the right proof," or at least it is the right style of proof. The theorem is combinatorial, and Euler gave a combinatorial proof. Euler employed the relationship among vertices, edges, and faces very directly. When a vertex was removed, faces and edges were added or removed to compensate, and doing so did not change the alternating sum.

By contrast, Legendre introduced seemingly unrelated concepts—spheres, angles, and area—to prove the theorem. His approach is valid and very clever, but it does not illustrate *why* the theorem is true—at least not in a transparent way. Nevertheless, Legendre's proof gave the first indication that there is more here than simply a combinatorial theorem. The fact that we *can* prove the theorem using metric geometry suggests an important relationship between Euler's formula and geometry. We will return to this topic—the relationship between Euler's formula and geometry—in chapters 20 and 21.

We make one last remark about Legendre's proof. With Euler's approach we were careful (more careful than Euler was himself) to apply the formula only to convex polyhedra. Like Euler, Legendre assumed that his polyhedra were convex. However, in the appendix to a paper written in 1809, Louis Poinsot (1777–1859) made the observation that Legendre's proof applied to a slightly more general class of polyhedra than the convex ones, a class of polyhedra called *star-convex*.[6]

The first step in Legendre's proof was to project the polyhedron onto the sphere. To accomplish this we need an interior point x from which

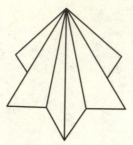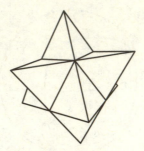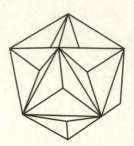

Figure 10.12. Star-convex polyhedra.

Figure 10.13. Louis Poinsot.

the projection can be made. This point must have the property that it can "see" every point in the polyhedron. For a convex polyhedron we can chose x to be any interior point. Most nonconvex polyhedra, however, have no such central point, but those that do are called star-convex. Kepler's star polyhedra, shown in figure 6.6, are examples of star-convex polyhedra, as are the ones shown in figure 10.12. For each of these, there is an interior point that is "all seeing," and from this point the projection can be made. Poinsot explained it as follows.

> [Euler's formula] still endures for any polyhedron with reentrant solid angles, provided that one can find, in the inside of the solid, a point which is the center of a sphere such that when the faces of the solid are cast there by lines that lead to the center, there is no duplication of these projections on the sphere; I mean to say, provided no face, in part or as a whole, is cast on the projection of another one; which applies, as one sees, for an infinite number of

polyhedra with reentrant solid angles. One recognizes the truth of this proposition easily from the very proof of Mr. Legendre, in which there will be nothing to change.[7]

Thanks to Legendre, by the end of the nineteenth century Euler's formula was put on sound footing for all convex polyhedra, and his popular textbook spread the beauty of Euler's formula to a broad audience. In the years that followed, Poinsot and other important mathematicians became entranced by this elegant relation. They searched for new proofs and for further generalizations. To understand some of these generalizations, we need to investigate the field of graph theory. The origins of this field date back, not surprisingly, to Euler—and a recreational mathematics puzzle from the city of Königsberg.

CHAPTER 11

A STROLL THROUGH KÖNIGSBERG

What is the use of going right over the old track again?
There is an adder in the path which your own feet have worn.
You must make tracks into the Unknown.
—Henry David Thoreau[1]

In order to place Euler's formula in a modern context, we must discuss a mathematical field called *graph theory*. This is not the study of graphs of functions that we encountered in high school precalculus ($y = mx + b$ is a line, $y = x^2$ is a parabola, and so on.). It is the study of graphs such as those shown in figure 11.1. They are made of points, called *vertices*, and lines joining these points, called *edges*.*

In 1736, during his first stay in St. Petersburg, Euler tackled the now famous problem of the seven bridges of Königsberg. His contribution to this problem is often cited as the birth of graph theory and topology.

The city of Königsberg was founded by the Teutonic Knights in 1254. At that time it was located in Prussia, near the Baltic sea, on a fork in the River Pregel. Later it became the capital of East Prussia. The city, which was heavily damaged by Allied bombing during World War II, fell under Soviet control following the Potsdam agreement. There were many changes in Königsberg after it became a Soviet state—most of the native Germans were expelled, the name of the city was changed to Kaliningrad, and the river was renamed the Pregolya. Today Kaliningrad is part of Russia and is the capital of the Kaliningrad Oblast region. Kaliningrad Oblast has the unique distinction that it is not connected to the rest of Russia; it is surrounded by Poland, Lithuania, and the Baltic Sea. Unlike cities such as Stalingrad and Leningrad, Kaliningrad has not reverted to its

---

*Sometimes graphs are called *networks*, with the vertices and edges called *nodes* and *links*, respectively.

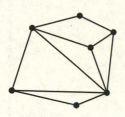 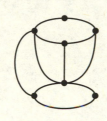 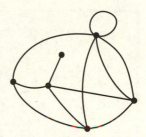

Figure 11.1. Graphs.

pre-Communist name. The most famous resident of Königsberg was the eighteenth century philosopher Immanuel Kant (1724–1804). Also from Königsberg came Christian Goldbach, the mathematician to whom Euler announced the discovery of his polyhedron formula.

The city is located on a fork in the river, and sitting in the middle of the river, near the fork, is Kneiphof Island. In Euler's time, there were seven bridges crossing the river joining the various banks and the island (see figure 11.2). As the story goes, the residents of Königsberg would leisurely walk around their city and entertain themselves by attempting to cross each of the seven bridges exactly once. No one was able to find such a route. This supposed pastime became the bridges of Königsberg problem:

> *Is it possible for a pedestrian to walk across all seven*
> *bridges in Königsberg without crossing any bridge twice?*

It is not known how Euler learned of this problem. Perhaps he heard it from his friend Carl Ehler, the mayor of Danzig, Prussia, who corresponded with Euler on behalf of a local professor of mathematics. We have letters between Ehler and Euler during the period 1735–1742, some of which discuss the Königsberg bridge problem. We do know that initially Euler was indifferent. In 1736, in a letter to Ehler, Euler wrote:

> Thus you see, most noble Sir, how this type of solution bears little
> relationship to mathematics, and I do not understand why you
> expect a mathematician to produce it, rather than anyone else, for
> the solution is based on reason alone, and its discovery does not
> depend on any mathematical principle.[2]

Eventually, Euler spent time thinking about the problem. The same feature that at first turned him off eventually piqued his interest: the problem did not fit comfortably within the existing mathematical framework. He

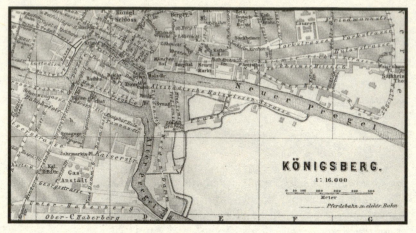

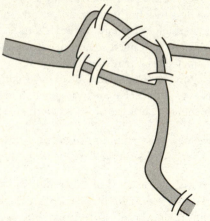

Figure 11.2. The seven bridges of Königsberg.

realized that the matter seemed geometrical, yet there was no need for exact distances. Information about relative positions was all that was needed.

Another letter from 1736, this one written to the Italian mathematician and engineer Giovanni Marinoni (1670–1755), said:

> This question is so banal, but seemed to me worthy of attention in that geometry, nor algebra, nor even the art of counting was sufficient to solve it. In view of this, it occurred to me to wonder whether it belonged to the geometry of position, which Leibniz had once so much longed for.[3]

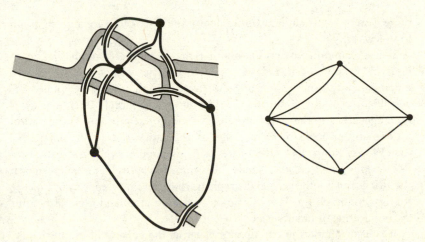

Figure 11.3. The graph associated with the bridges of Königsberg problem.

In this letter Euler used a term coined by Leibniz, *geometriam situs*, which translates to geometry of position. Later this term would become *anlaysis situs* (analysis of position), and eventually, topology. Leibniz was referring to a new field in mathematics, one that "deals directly with position, as algebra deals with magnitudes."[4] There is some disagreement among scholars whether Euler misunderstood Leibniz's use of this term; nevertheless, Euler agreed with Leibniz's recognition of the need for new mathematical techniques to handle this situation.

In 1736 Euler presented his paper "Solutio problematis ad geometriam situs pertinentis" ("The solution of a problem relating to the geometry of position") to the St. Petersburg Academy.[5] It was published in 1741. In it Euler solved the Königsberg bridge problem and, in his typical style, generalized his solution to any layout.

Euler realized that the only important details in the problem are the relative locations of the land masses and the bridges joining them. Using a diagram, we can abstract the situation easily and elegantly. Place a vertex on each piece of land (one on each of the three banks and one on the island), and join each pair of vertices by as many edges as there are bridges connecting the landmasses. The resulting graph is shown in figure 11.3.

In this way, we reduce the problem to one about a graph—i.e., is it possible to trace this graph with a pencil without lifting the pencil and without redrawing any edge? From this example we can formulate the more general question: how do we determine if we can trace a given graph in this way?

It is a common misconception that the Königsberg graph, shown in figure 11.3, is found in Euler's paper. In reality, neither the Königsberg graph nor any other graph appears there. Graph tracing developed independently of the Königsberg bridge problem. Graph tracing puzzles first appeared in the early nineteenth century, both in mathematical articles and in books of recreational mathematics. It was not until 1892 that W. W. Rouse Ball (1850–1925), in his popular work *Mathematical Recreations and Problems,*[6] made the connection between Euler's result on the bridges of Königsberg and graph tracing. The first appearance of the Königsberg graph was in Ball's book, over a hundred and fifty years after the publication of Euler's paper.

It is also common to cite Euler's paper as the genesis of graph theory. This attribution is not unreasonable. Although Euler never drew a graph in his paper, his abstract treatment of the problem resembles the graph theory argument. His application of *geometriam situs*, what would later be called topology, to the problem and his recognition of the novelty of this method signals the start of this new field.

In order to discuss his solution we need a few definitions. As with polyhedra, the *degree* of a vertex is the number of edges emanating from it. If there is a *loop* at this vertex (an edge starting and ending at the vertex, such as in the right graph in figure 11.1), then the loop contributes two to the degree. The graph for the bridges of Königsberg problem has three vertices of degree 3 and one vertex of degree 5. A graph is *connected* if it is possible to get from any vertex to any other vertex by following a sequence of edges.

A tracing of a graph that begins at one vertex and ends at another vertex is called a *walk*. We are interested in a very special class of walks, ones that visit each edge exactly one time: this is called an *Euler walk*. If the Euler walk begins and ends at the same vertex, then it is called an *Euler circuit*. In general, a *circuit* is a walk along a graph that starts and ends at the same vertex and never visits the same edge twice. A (non-Eulerian) circuit may not visit every edge.

Using the language of graph theory, we recast the bridges of Königsberg problem as follows.

> Does the bridges of Königsberg graph (figure 11.3) have an Euler walk? More generally, how do we determine if a graph has an Euler walk?

Euler solved both of these problems. Translated in to modern language, it is stated as follows.

> A graph has an Euler walk precisely when it is connected and there are zero or two vertices of odd degree. If there is a pair of vertices of odd degree, then the walk must start at one of these vertices; otherwise the walk may begin anywhere.

Using these criteria we easily solve the bridges of Königsberg problem. Because the graph has four vertices of odd degree, there is no Euler walk! It is no wonder that the residents of Königsberg were so frustrated in their search for the ideal afternoon stroll.

But why is Euler's solution true? The requirement that the graph be connected is obvious. The insistence that the graph have zero or two vertices of odd degree requires some thought. To prove the theorem, we have two objectives. First we must show that any graph with an Euler walk must have zero or two vertices of odd degree. Then we must show the converse: if a connected graph has zero or two vertices of odd degree, then it has an Euler walk.

Suppose we have a graph with an Euler walk; we will show that it must have zero or two vertices of odd degree. Place a sheet of tracing paper on top of the graph and proceed to trace the Euler walk. As we begin tracing the Euler walk, the first vertex will have degree one, and all other vertices will have degree zero. As we get to, and trace through, the second vertex, it will have degree two. From then on, each time we pass a vertex, the degree increases by two. This continues until we reach the end of the walk. At this point, we add one to the degree of the last vertex. If the walk starts and ends at two different vertices, then these two vertices will have odd degree, and they will be the only vertices of odd degree. If the walk starts and ends at the same vertex, then it, and all other vertices, will have even degree.

Euler took the converse for granted: if a graph has zero or two vertices of odd degree, then it has an Euler walk. The first demonstration of this fact was given by Carl Hierholzer (1840–1871), and was published posthumously in 1873.[7]

Begin with a connected graph having either zero or two vertices of odd degree. If the graph has a pair of vertices of odd degree, then place the pencil at one of these vertices; otherwise put it at any vertex. Begin tracing in any direction. Upon reaching the first vertex, choose randomly a new edge to follow. Continue in this way, making arbitrary choices at each vertex (while

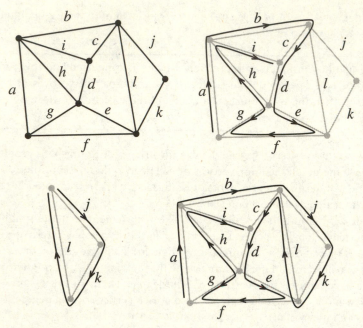

Figure 11.4. Building an Euler walk.

avoiding edges that were previously visited, of course), until it is impossible to go any farther. By the argument given earlier, if we began at a vertex of odd degree, then the end of this tracing will be at the other vertex of odd degree; otherwise the tracing will end at the starting vertex. In figure 11.4 path $abcdefghi$ is such a walk.

If this path does not pass through every edge in the graph, then remove all of the traced edges and look at the remaining graph (it may no longer be connected). Place your pencil at a vertex that was in your original tracing. As before, trace this graph until it is impossible to trace any farther. In our example, we obtain the walk jkl. Now insert this new tracing into the appropriate location of the walk that you constructed previously. In our example, we may insert jkl between edges b and c of the original walk. So, we obtain $abjklcdefghi$, which is an Euler walk. In general, it may be necessary to make several such insertions before all edges are traced.

Notice that we learned more about graph tracing than is evident in the solution given above. Our discussion was aimed at finding Euler walks, but we also determined when the walk can begin and end at the same vertex.

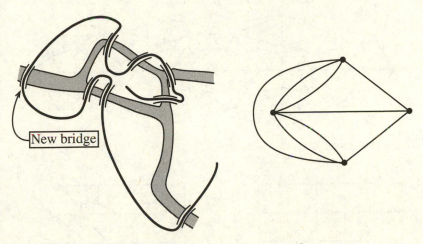

Figure 11.5. A new bridge in Königsberg and the new graph.

That is:

> A graph has an Euler circuit precisely when it is connected and has no vertices of odd degree. In this case, the Euler circuit can begin and end at any vertex.

In 1875, a century and a half after Euler analyzed the walking routes of the city of Königsberg, the city built a new bridge.[8] It was erected west of Kneiphof Island from the northern bank to the southern bank (see figure 11.5). With this in place, the residents of Königsberg could finally take a stroll across all the bridges and visit each bridge exactly one time, for there were now exactly two vertices of odd degree—the vertices corresponding to the island and the land between the fork. Of course, some of the townsfolk were not able to begin their walk at their front doorstep, and no one was able to end his or her walk where it began.

This solution to the Königsberg bridge problem illustrates a general mathematical phenomenon. When examining a problem, we may be overwhelmed by extraneous information. A good problem-solving technique strips away irrelevant information and focuses on the essence of the situation. In this case details such as the exact positions of the bridges and land masses, the width of the river, and the shape of the island were extraneous. Euler turned the problem into one that is simple to state in graph theory terms. Such is the sign of genius.

Figure 11.6. Listing's graph-tracing puzzle.

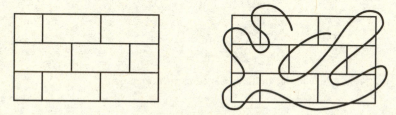

Figure 11.7. An incorrect solution to the brick wall puzzle.

We conclude with three examples. In 1847, Johann Benedict Listing (1808–1882), a mathematician whom will meet again later, produced the graph shown in figure 11.6 to illustrate the tracing problem (we draw the graph as Listing did and omit the vertices at the intersections).[9] Does it have an Euler walk? Does it have an Euler circuit? The reader may wish to think about this problem before proceeding.

We see that every vertex has even degree except for the left-most and right-most—these vertices have degree five. Because there are exactly two vertices of odd degree, Listing's graph does have an Euler walk, and every Euler walk must begin at one of these vertices and end at the other. Because the graph has vertices of odd degree, there is no Euler circuit.

The second example is a variation on the bridge problem. Consider the drawing resembling a brick wall shown in figure 11.7. Is it possible to draw a single unbroken curve that crosses each of the line segments in the figure exactly one time (the curve may begin and end in different bricks)? The attempt shown on the right is not a valid solution because there is one uncrossed segment.

It is not possible. We can justify this claim by transforming the problem into a graph-tracing problem. Place one vertex inside each brick and one vertex outside the figure. Draw one edge from a vertex to another vertex for each segment separating the corresponding bricks in the original picture

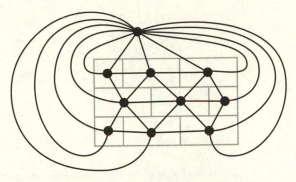

Figure 11.8. A graph associated with the brick wall puzzle.

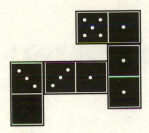

Figure 11.9. A typical game of dominoes.

(see figure 11.8). It suffices to determine whether this graph has an Euler walk. Because the graph has four vertices of degree five, it has no Euler walk. So there is no curve with the desired properties.

Finally, we apply graph theory and Euler walks to the game of dominoes. This example was concocted by Orly Terquem (1782–1862) in 1849.[10] In a standard set of dominoes, each half of a domino has zero to six pips. No two dominoes in the set are alike, and every combination is present in the set. This gives a total of 28 dominoes. Play alternates as each player lays down a domino in such a way that the number on half of her domino abuts the same number on an existing domino. A domino with the same number of pips on each half can be placed in a T formation against a tile with that number of pips on one half (see figure 11.9). Play ends when a player is unable to lay down another domino. We ask, will a game always end with a player holding dominoes in his cache? Or is it possible to lay down all of the dominoes and never get stuck?

To analyze this problem we create a graph as follows. Start with seven vertices labeled 0 through 6. Each domino corresponds to an edge on this

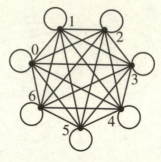

Figure 11.10. The graph corresponding to the collection of all domino tiles.

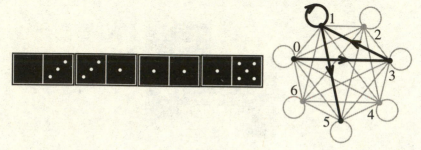

Figure 11.11. Part of a game of dominoes in which every tile is played, and the corresponding graph.

graph. A domino with m pips on one half and n pips on the other becomes an edge from vertex m to vertex n. Putting all of the dominoes on the graph we obtain figure 11.10. Notice that there is a loop at each vertex corresponding to dominoes with the same number of pips on each half.

Each vertex in the domino graph has degree eight. Because the degree of every vertex is even, this graph has an Euler walk. So, we can trace the entire graph passing through each edge exactly one time. This observation is the key to answering our question. To show that we can play all of the dominoes, it suffices to find one configuration that achieves this end. The one we produce is simple (although it would be unlikely to arise in actual play)—a line of dominoes.

Start with the first edge in the Euler walk. Suppose it joins vertices 0 and 3. Lay down the domino containing zero pips and three pips. Now consider the second edge in the walk. We know that this edge must begin at vertex 3. Suppose the edge joins vertices 3 and 1. We then lay down the domino with three pips and one pip on the end of the previous domino (see figure 11.11).

We continue on in this fashion, laying down tiles as we go. Because we are following an Euler walk, we will get each edge exactly one time. Thus we will be able to lay down every domino in the set.

As these examples show, graph theory has some wonderful applications to recreational mathematics. However, it is also a very important field of mathematics that has numerous practical applications in such diverse areas as computer science, networking, social structures, transportation systems, and epidemiological modeling. We will see graph theory again in the chapters that follow. In particular, we will create an analogue of Euler's formula for a certain class of graphs.

CHAPTER 12

CAUCHY'S FLATTENED POLYHEDRA

Cauchy is mad and there is nothing that can be done about him, although, right now, he is the only one who knows how mathematics should be done.
—Niels Abel[1]

In the hundred years after Euler's proof of his polyhedron formula, there were many new proofs and a variety of generalizations to exotic polyhedral shapes. The first significant generalization came from Augustin-Louis Cauchy, who also gave an ingenious new proof.

Cauchy was born in Paris in 1789. He was the eldest son of a senior government administrator. Although his family was displaced during the Reign of Terror, his father saw that he received a good education. As a youth he became acquainted with the mathematicians Pierre-Simon Laplace (1749–1827) and Joseph-Louis Lagrange, and the chemist Claude Louis Berthollet (1748–1822), so he was exposed early in his life to influential scholars.

Cauchy had a brief career as a military engineer working on the Ourcq Canal, the Saint-Cloud Bridge, and the Cherbourg naval base. His first mathematical publications were composed in 1811, two years before before he returned to Paris to begin a mathematical career. In 1815 he began his employment at the École Polytechnique.

Cauchy was extremely prolific. His voluminous output was second only to Euler's; his collected works, which include at least seven books and eight hundred papers, fill twenty-seven sizable volumes. Although it is likely apocrypha, it was said that a newly-instituted rule at the French Academy limiting the number of publications per contributor per year was in response to the deluge of publications flowing from Cauchy's pen.

Cauchy made deep and substantial contributions to many areas of mathematics, including complex analysis, real analysis, algebra, differential

Figure 12.1. Augustin-Louis Cauchy.

equations, probability, determinants, and mathematical physics. He was one of the early champions of the need for rigor in mathematics. Many of the fundamental ideas of calculus introduced by Newton, Leibniz, Euler, and others were finally placed on a firm theoretical foundation by Cauchy. We can thank him for what are essentially the modern definitions of continuity, limit, derivative, and definite integral. Through his frequent lectures at the École Polytechnique and his numerous publications, his was an ever-present voice in the mathematical community for the first half of the nineteenth century.

It is a testament to Cauchy's influence that there are so many theorems, properties, and concepts named after him—perhaps more than for any other mathematician, even Euler. Yet it seems that Cauchy became one of the great mathematicians in spite of himself. He often published great works without seeming to recognize their depth and importance. The mathematician Hans Freudenthal (1905–1990) wrote, "In nearly all cases he left the final form of his discoveries to the next generation. In all that Cauchy achieved there is an unusual lack of profundity ...He was the most superficial of the great mathematicians, the one who had a sure feeling for what was simple and fundamental without realizing it."[2]

Although Cauchy was greatly admired as a mathematician, he was not usually admired as a person. He was known to be stubborn, and he had a flair for melodrama. These traits are typified by a self-imposed exile from France to Turin and Prague that lasted almost a decade. He was a political conservative who followed the ousted Bourbon king Charles X

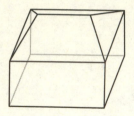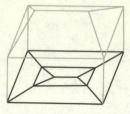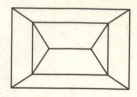

Figure 12.2. Cauchy projected the polyhedron into the bottom face.

after the July Revolution of 1830. Before he left France and after he returned, he refused to swear the required oath of allegiance to the new regime or even to agree not to speak out. He was a staunch Catholic whose charitable actions were overshadowed by behavior that betrayed him as a "bigoted, selfish, narrow-minded fanatic."[3] One biographer wrote that Cauchy was "an arrogant royalist in politics and a self-righteous, preaching, pious believer in religion ... most of his fellow scientists disliked him and considered him a smug hypocrite."[4]

Cauchy wrote his first mathematical papers while he was still an engineer. These papers contain his results on polyhedra, including his rigidity theorem (which we discussed in chapter 5) and his work on Euler's formula. These important results were among the very few contributions Cauchy made to geometry.

The first notable feature that distinguishes Cauchy's proof of Euler's formula from those of his predecessors is that the polyhedra in his proof are hollow, not solid. More specifically, he considers the "convex surface of a polyhedron."[5] Because of this language and because elsewhere in the paper he carves a polyhedron into smaller polyhedra, it appears that he still views a polyhedron as solid, and only for the purpose of the proof does he assume that it is hollow.

The first step in Cauchy's proof is to transform this hollow polyhedron into a graph in the plane. He removes a face from the polyhedron, then "by transporting onto this face all the other vertices without changing their number, one will obtain a planar figure made up of several polygons contained in a given contour." Cauchy clarifies this instruction by saying that "the remaining faces ... could be regarded as forming a suite of polygons contained in the outline of the removed face." We see this process illustrated in figure 12.2, where the house-shaped polyhedron is transported onto the ground floor.

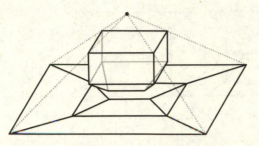

Figure 12.3. The flattened polyhedron viewed as a shadow of the edges.

Joseph Diaz Gergonne (1771–1859), a contemporary of Cauchy's whom we will meet in chapter 15, described this process as follows:

> Take a polyhedron, one of its faces being transparent; and imagine that the eye approaches this face from the outside so closely that it can perceive the inside of all the other faces; this is always possible when the polyhedron is convex. The things being so arranged, let us imagine that on the plane of the transparent face a perspective is made of the set of all the others.[6]

In his wonderful book *Proofs and Refutations*, Imre Lakatos (1922–1974) put a modern spin on Gergonne's idea by suggesting that a camera be placed near the removed face, and the inside of the polyhedron be photographed. The graph would then appear on the photographic print. Yet another common way of visualizing this flattened polyhedron is as the shadow of the edges when a light bulb is situated near the removed face (figure 12.3).

Cauchy realized that it sufficed to understand the relative numbers of vertices, edges, and faces in the graph. Cauchy proved that every such graph satisfies $V - E + F = 1$. Once he established this fact, it was easy to complete the proof of the polyhedron formula. The graph obtained from transporting the polyhedron onto the plane has the same number of edges and vertices as the polyhedron, but it has one fewer face. Because $V - E + F = 1$ for the graph, $V - E + F = 2$ for the polyhedron. Cauchy's adaptation of Euler's formula to graphs in the plane is one of the most useful generalizations.

The idea of Cauchy's proof is to add and remove edges in such a way that at each step the quantity $V - E + F$ does not change. Then in the end, we obtain a single triangle, which satisfies $V - E + F = 3 - 3 + 1 = 1$, implying that $V - E + F = 1$ for the original graph. The first step in

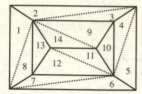 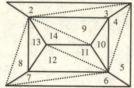 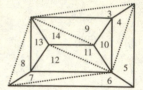

Figure 12.4. The order of triangle removal from the triangulated graph.

Cauchy's proof is to divide the graph into triangular regions by adding diagonals to all nontriangular faces (see figure 12.4). This procedure is called *triangulating* a graph. Each time a diagonal is added, the number of edges increases by one, the number of faces increases by one, and the number of vertices remains the same. Thus $V - E + F$ is the same for the original graph as for the modified graph. Once the graph has been triangulated, we decompose it by removing triangles from the outside, one at a time, until a single triangle remains (one possible order is given by the labels in figure 12.4).

Notice that a triangle on the outside of the graph has either one or two exterior edges. In the first case the triangle can be removed by taking away the single edge and no vertices (such as the removal of triangle number 1). In the second case the triangle can be removed by taking away two edges and a vertex (such as the removal of triangle number 2). In either case, the quantity $V - E + F$ remains unchanged. Thus, it must be the case that $V - E + F = 1$ for the original graph.

Cauchy's proof was later criticized. Just as Euler ran into trouble by failing to give explicit instructions on what order to remove the pyramids, Cauchy did not give reliable instructions on how to cut away the triangles. If we are not careful, it is possible to follow Cauchy's algorithm and obtain a disconnected graph for which the relation fails to hold. For instance, in figure 12.5 we removed triangles in the wrong order and obtained a disconnected graph that fails to satisfy Euler's formula ($V = 10$, $E = 14$, and $F = 6$). Nonetheless, it is always possible to use Cauchy's technique to decompose a graph without this situation occurring.

As we have already mentioned, Cauchy had a track record for proving theorems without realizing their importance or without following them to their logical conclusions. His proof of Euler's polyhedron formula is a perfect example of this. In his paper he makes the explicit assertion that his proof applies to convex polyhedra. Indeed it does, but it actually applies to a much more general class of polyhedra. The key step in Cauchy's proof is

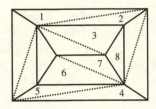 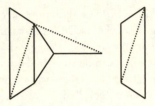

Figure 12.5. Cauchy's method can yield a degenerate polygon.

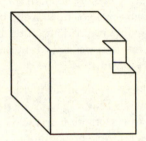 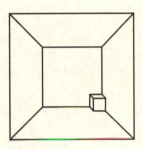

Figure 12.6. A cube with an indented corner and its graph.

removing a face and transporting the rest of the polyhedron onto the plane in such a way that none of the faces overlap. This can be accomplished for any convex polyhedron, but it can also be accomplished for many other polyhedra.

For instance, Cauchy's proof applies without modification to the nonconvex polyhedron shown in figure 12.6. To see this, simply place Lakatos's camera near the bottom face of the cube.

Lakatos and the mathematician Ernst Steinitz (1871–1928) claim that Cauchy knew his proof applied to some or perhaps all nonconvex polyhedra, but this is not clear from what Cauchy wrote. The discrepancy arises from Cauchy's casual use of the word "convex." It did not appear in the statement of the theorem, but in the proof he said he was considering "the convex surface of a polyhedron." He never addressed the discrepancy, so it is impossible to guess what he did and did not know.

Regardless of whether Cauchy recognized that the result could be extended to some nonconvex polyhedra, it was quickly seen by others. In 1813, the same year that Cauchy's paper was published, Gergonne gave his own proof of Euler's formula. Afterward he wrote, "One might prefer still, with reason, the beautiful proof of Mr. Cauchy, who has the precious advantage of not assuming that the polyhedron is convex."[7]

With a little imagination we can apply Cauchy's proof to a still broader class of polyhedra. Modern versions of Cauchy's proof describe a polyhedron as made out of rubber. If a face can be removed and the rest of the polyhedron can be stretched out onto the plane without any overlapping or folding, then Cauchy's proof applies. In chapter 15 we will see pathological examples of polyhedra that do not have this property—when a face is removed, the rest of the polyhedron cannot be transported to the plane. The key property, it turns out, is that the polyhedron is "sphere-shaped." We will discuss this seemingly vague property in more detail in chapter 16. Cauchy was agonizingly close to recognizing this essential property. Had he noticed this, he would have made a significant contribution to the yet-unformed field of topology, or *analysis situs* as it was known in the early days. As Jacques Hadamard (1865–1963) wrote in 1907:

> I consider to be one of the most remarkable events in the history of science, the error which Cauchy made while believing to prove the theorem of Euler without introducing any assumption on the nature of the studied polyhedron. It is indeed a principle of paramount importance which eluded him and which he left for Riemann to discover: the fundamental role of the *analysis situs* in mathematics.[8]

Just as Cauchy missed the full strength of his proof for polyhedra, he also did not recognize the full strength of this proof for graphs. For example, it was Arthur Cayley (1821–1895) who, in 1861, observed that Cauchy's proof applied equally well to graphs that had curved edges (this fact was noticed independently by Listing in 1861 and Camille Jordan [1838–1922] in 1866).[9]

In the statement of his theorem Cauchy assumed that the graph is a collection of polygons contained in a polygonal outline. As we will see in the next chapter, we can give a much more general statement about graphs, but in order to do so we must introduce some modern terminology.

CHAPTER 13

PLANAR GRAPHS, GEOBOARDS, AND BRUSSELS SPROUTS

In most sciences one generation tears down what another has built, and what one has established, another undoes. In mathematics alone each generation adds a new story to the old structure.
—Hermann Hankel[1]

In the previous chapter we saw Cauchy's clever technique for proving Euler's formula. He took a polyhedron, removed a face, and projected the rest down onto the plane. Then he proved that $V - E + F = 1$ for this figure, so $V - E + F = 2$ for the polyhedron. The connection to graph theory should be obvious. At first glance it appears that it would be trivial to generalize Euler's formula to graphs that are not projections of polyhedra and which may possess edges that are curved.

The difficulty with extending Euler's formula is that it does not apply to every graph. Counting vertices and edges is easy—they are the building blocks from which graphs are made—but a graph need not have faces. Even when a graph is drawn on paper, the edges may not divide the region into faces. For example, the edge PR in the left-hand graph of figure 13.1 crosses edge QS, so it cannot be the boundary of a face. However, the graph can be redrawn (as on the right) without any crossings, thus dividing the region into faces. A graph that can be drawn without edges crossing is called *planar*.

On a polyhedron, a face is a region bounded by a polygon. For graphs, we adopt a looser definition. A face can be bounded by one edge, like the loop from P to P in figure 13.1. A face can be bounded by two edges, such as the pair of edges from Q to R. (A pair of edges that connect the same two vertices is called *parallel*.) A face may even have an edge protruding into its interior, such as the edge from S to T.

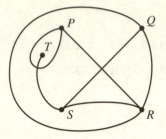
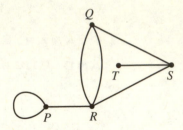

Figure 13.1. Two representations of the same graph.

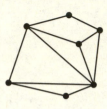
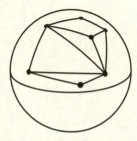

Figure 13.2. Placing a planar graph on a sphere.

Many graph theorists count the exterior region as a face. If the graph is viewed as an island, then this unbounded face is the sea stretching off to infinity in every direction. Although it is somewhat inelegant to call this unbounded region a face, it is often more useful to include it than to exclude it. One way to reconcile this problem is to envision the graph as an island not on an unending sea, but on a globe (figure 13.2). In this way the unbounded face becomes finite.

So we have the following generalization of Euler's formula for planar graphs.

EULER'S FORMULA FOR PLANAR GRAPHS
A connected planar graph with V vertices, E edges, and F faces satisfies $V - E + F = 2$.

If we do not count the unbounded region as a face, then Euler's formula is $V - E + F = 1$. Notice that the graph in figure 13.1 has 5 vertices, 7 edges, and 4 faces, and $5 - 7 + 4 = 2$, as expected.

As an elementary example, consider a *tree*. A tree is a connected graph that has no circuits (see figure 13.3). Because a tree has no circuits, the

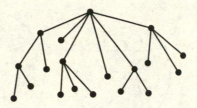

Figure 13.3. A tree.

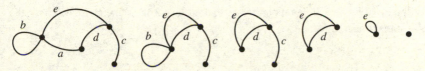

Figure 13.4. Reducing a planar graph to a single vertex by removing edges a, b, c, d, then e.

only face is the unbounded one, so Euler's formula yields $V - E + 1 = 2$, or $V = E + 1$. In other words, the number of vertices in a tree exceeds the number of edges by one. The tree in figure 13.3 has 19 vertices and 18 edges.

There are many proofs of Euler's formula for graphs. We will give a short one in which, like Cauchy, we remove edges from the graph one at a time. But we will be careful to avoid his mistake.

Begin with any connected planar graph. Pick any edge. This edge either joins two different vertices or it is a loop from one vertex to itself. Suppose the edge joins two vertices. Then shrink the edge until it completely disappears and its two endpoints become one. This can be done so that the resulting graph is planar (see the shrinking of edges a, c, and d in figure 13.4). This procedure decreases the number of edges by one and the number of vertices by one, and it leaves the number of faces unchanged. Thus $V - E + F$ is unchanged. Now suppose that the edge is a loop. In this case, simply remove the edge from the graph (see the removal of edges b and e in figure 13.4). Doing so decreases the number of edges and faces by one, but leaves the number of vertices unchanged. Thus $V - E + F$ is unaltered.

Continue this process of edge removal until a single vertex remains. At this point we have one vertex, no edges, and one face (the exterior region). Thus $V - E + F = 2$. Because $V - E + F$ was unchanged throughout the process, $V - E + F = 2$ for the original graph.

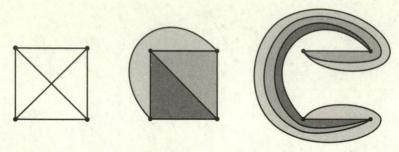

Figure 13.5. Two different planar representations of the same graph will have the same number of faces.

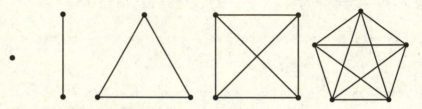

Figure 13.6. The complete graphs K_1, K_2, K_3, K_4, and K_5.

An interesting consequence of this formula is that for a planar graph with E edges and V vertices, every planar representation will have the same number of faces. In other words, if ten people took a planar graph, placed the vertices wherever they desired, and placed the edges on the graph in such a way that no two crossed, all of the graphs would have the same number of faces ($F = 1 + E - V$, excluding the unbounded face). For instance, in figure 13.5 we have a graph with four vertices and six edges, and we show two planar representations, both of which have three faces.

Because Euler's formula applies only to planar graphs, we can often use it to prove that a given graph is not planar. To illustrate this idea we introduce two important families of graphs: *complete graphs* and *complete bipartite graphs*.

The complete graph with n vertices, denoted K_n, has n vertices and exactly one edge between each pair of vertices. It is the largest possible graph with n vertices that has no loops or parallel edges. The graphs K_1 through K_5 are shown in figure 13.6. If we removed the loops from the domino graph (figure 11.11), we would have K_7.

A relative of the complete graph is the complete bipartite graph. A complete bipartite graph has the property that its vertices can be split into

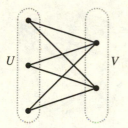
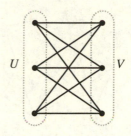

Figure 13.7. The complete bipartite graphs $K_{3,2}$ and $K_{3,3}$.

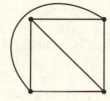

Figure 13.8. K_4 and $K_{3,2}$ are planar graphs.

two collections, U and V, such that there are no edges joining vertices in U to vertices in U, no edges joining vertices in V to vertices in V, and exactly one edge joining each vertex in U to each vertex in V. If U has m vertices and V has n vertices, we denote the resulting complete bipartite graph $K_{m,n}$. The graphs $K_{3,2}$ and $K_{3,3}$ are shown in figure 13.7. The typical example of a complete bipartite graph that is found in every introductory graph theory textbook is the utility company graph. The collection U consists of utility companies (gas, water, electric, etc.) and the collection V consists of the customers. Because every customer must have each utility, the graph obtained is a complete bipartite graph.

We would like to determine which complete graphs and which complete bipartite graphs are planar. It is easy to show that K_1, K_2, K_3, K_4, $K_{m,1}$, and $K_{m,2}$ are planar graphs. For example, in figure 13.8 we see that K_4 and $K_{3,2}$ are planar. It turns out that none of the rest are planar. Using Euler's formula we will prove that K_5 and $K_{3,3}$ are nonplanar.

To prove that K_5 is not planar we use a proof technique called *proof by contradiction*, or *reductio ad absurdum*. We assume the opposite of what we need to prove (we assume that K_5 is planar) and show that this leads to a logical contradiction. Then we are able to conclude that K_5 is not planar. G. H. Hardy wrote, "*Reductio ad absurdum*, which Euclid loved so much, is one of a mathematician's finest weapons. It is a far finer gambit than any

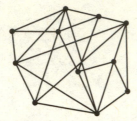 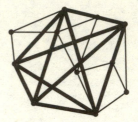

Figure 13.9. An example of a nonplanar graph containing K_5.

chess gambit: a chess player may offer the sacrifice of a pawn or even a piece, but a mathematician offers the game."[2]

Suppose K_5 is a planar graph. Then we can draw K_5 in the plane in such a way that no edges cross. K_5 has 5 vertices and 10 edges. Euler's formula for planar graphs states that $V - E + F = 2$, thus our planar drawing of K_5 must have 7 faces including the unbounded face (because $2 = 5 - 10 + F$).

Each edge borders two faces, so $2E = pF$ where p is the average number of sides on all faces. K_5 is a complete graph so it has no loops or parallel edges. Because there are no loops, there are no faces bounded by only one edge, and because there are no parallel edges, there are no faces bounded by two edges. Thus, the average number of edges per face is at least three. So, $p \geq 3$, and $2E \geq 3F$. But $F = 7$ and $E = 10$ implies that $20 \geq 21$, which is a contradiction. It must be the case that K_5 is not planar.

In a similar way we can prove that the complete bipartite graph $K_{3,3}$ is not planar (give it a try!). The key difference is that because $K_{3,3}$ is bipartite, a path that begins and ends at the same vertex must have an even number of edges. So there can be no three-sided faces, either.

In general, we have the following theorem.

K_n is not planar when $n \geq 5$ and $K_{m,n}$ is not planar when both $m \geq 3$ and $n \geq 3$.

It turns out that in some sense K_5 and $K_{3,3}$ are the only obstacles for a graph to be planar. A famous theorem called Kuratowski's reduction theorem states that the only way a graph can fail to be planar is if it contains a copy of K_5 or $K_{3,3}$ inside it. For instance, the graph shown in figure 13.9 is not planar because it contains a copy of K_5.

Next, we give another interesting application of Euler's formula called Pick's theorem. It was proved by Georg Alexander Pick (1859–c. 1943)

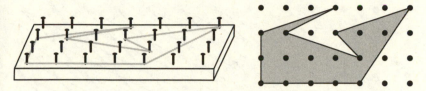

Figure 13.10. A polygon formed on a geoboard.

in 1899.[3] Pick was an Austrian mathematician who spent much of his life in Prague. He died in the Theresienstadt concentration camp in Czechoslovakia.

To introduce Pick's theorem we turn to the *geoboard*, a popular teaching tool invented by Caleb Gattegno (1911–1988) that gives children a hands-on way of learning basic geometry. A geoboard can be made at home by hammering nails halfway into a wooden board so that they form a square grid. The students can wrap rubber bands around the nails to form polygons (see figure 13.10). The teacher can then discuss geometric concepts such as perimeter, angle, area, and the Pythagorean theorem.

Pick's theorem gives a very easy way to compute areas of even the most complicated nonconvex polygons (we do insist, however, that the rubber band not cross itself).

PICK'S THEOREM

If there are B nails on the boundary of the polygon and I nails in the interior of the polygon, then the area is $A = I + B/2 - 1$.

For instance, because the polygon in figure 13.10 has $B = 12$ and $I = 5$, the area is $5 + 12/2 - 1 = 10$.

It turns out that at least one forester in Oregon used an approximation of Pick's theorem to estimate the size of his forest.[4] The forester took a transparency with a lattice pattern of dots on it and placed it on a polygonal map of his land. To estimate the area he added the number of dots in the interior to half the number on the boundary (very nearly Pick's theorem!) and multiplied this by the appropriate scaling factor.

Pick's theorem is an elementary consequence of Euler's formula once we know the area of a *primitive* triangle. A triangle is primitive if it has no nails in its interior and boundary nails only at the vertices (such as the shaded triangle in figure 13.11). In other words, a triangle is primitive if $B = 3$ and $I = 0$. Surprisingly, every primitive triangle has area 1/2.

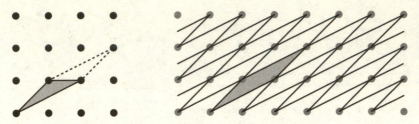

Figure 13.11. The parallelogram formed from a primitive triangle tiles the plane.

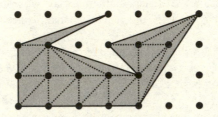

Figure 13.12. The polygon is triangulated into primitive triangles.

Unfortunately, this fact is somewhat tedious to prove. Instead of giving the proof we will give an indication of why it is true. It is easy to see the plane (an infinite geoboard) can be tiled by 1×1 squares. These tiles have the property that if they are shifted one unit up, down, right, or left, they will land on another square. Likewise, take the primitive triangle and extend it to form a parallelogram with double the area (see figure 13.11). Like the square, the parallelogram can tile the plane by repeatedly moving it one unit to the left, to the right, up, and down. So, like the square, the parallelogram has area 1. Thus the triangle has area 1/2.

We are now able to prove Pick's theorem. First, triangulate the polygon into T primitive triangles (such as in figure 13.12). If we count the unbounded region as a face, then $F = T + 1$. Because every triangular face has area 1/2, the total area of the polygon is $A = (1/2)T$.

Every bounded face has three sides, so the quantity $3T$ counts each edge twice except for the edges along the boundary, which are counted once. Because the number of edges on the boundary is the same as the number of vertices on the boundary, we have

$$3T = 2E - B,$$

or, solving for E,

$$E = \frac{3T}{2} + \frac{B}{2}.$$

The number of vertices is $V = I + B$. Applying Euler's formula we have,

$$2 = V - E + F$$

$$2 = (I + B) - \left(\frac{3T}{2} + \frac{B}{2}\right) + (T + 1)$$

$$2 = -\frac{T}{2} + \frac{B}{2} + I + 1.$$

So, the number of triangular faces is

$$T = 2I + B - 2,$$

and the total area is given by

$$A = \frac{T}{2} = \frac{1}{2}(2I + B - 2) = I + \frac{B}{2} - 1,$$

as claimed.

We end with two pencil-and-paper games. Despite their similarities, one is an intellectual challenge and the other a hoax on the players with the outcome determined before the first move.

According to Martin Gardner (b. 1914), the long-time mathematics columnist for *Scientific American*, the game of "sprouts" was invented at teatime one February afternoon in 1967 by John Horton Conway (b. 1937) and Michael Paterson at the University of Cambridge. It became an underground sensation. Conway wrote to Gardner that "the day after sprouts sprouted, it seemed that everyone was playing it. At coffee or tea times, there were little groups of people peering over ridiculous to fantastic sprout positions."[5]

The game begins with any number of dots on a blank sheet of paper. The first player draws a curve from one dot to another dot or back to itself, then places a new dot somewhere along the curve. Play alternates with both players drawing a curve from dot to dot and placing a new dot along the way. The only rules are that a new curve cannot pass through an existing curve or dot and that at most three lines can meet at any dot. The winner is the last player able to make a move. Figure 13.13 shows a 2-dot game in which player 2 wins in the fourth move.

The more dots at the outset, the longer the game, but the game cannot go on forever. At the start of an n-dot game there are $3n$ available places to attach edges. Each time a new edge is drawn, the number of open spots decreases by one (two are used up, one is added). So the longest the game could possibly last is $3n - 1$ moves.

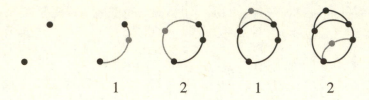

1   2   1   2

Figure 13.13. Player 2 wins this game of sprouts.

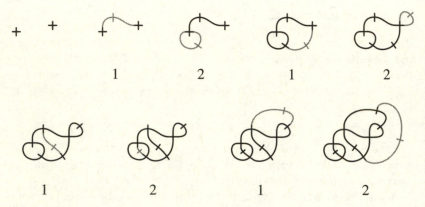

1   2   1   2

1   2   1   2

Figure 13.14. Player 2 wins this game of Brussels sprouts.

As it turns out, for small values of n there is a slight advantage to going either first or second, depending on the number of dots. If the game starts with 2 dots, the second player can always make choices that guarantees a win. The same is true for $n = 1, 6, 7,$ and 8 dots. On the other hand, the first player has the advantage for $n = 3, 4, 5, 9, 10,$ or 11 dots. Most of these cases ($n > 6$) were checked by computer.[6] For larger n values it is not known if either player has an advantage. Although there is a theoretical unfairness in this game, for all but the smallest of n-values the winning strategy is unknown. So in practice, it is still an intellectual challenge to win the game.

Later Conway invented a variant of sprouts that he dubbed "Brussels sprouts." Instead of starting with n dots, the game begins with n plus-shaped crosses. Each turn a player joins two free arms with a curve and puts a new plus sign (producing two free arms) in the middle of the edge (see figure 13.14). Unlike sprouts, the rules for Brussels sprouts allow four edges to meet at the same location. As before, the winner is the last player able to make a move. It turns out that the humorous name Brussels sprouts was a tip-off that the game itself was a joke.

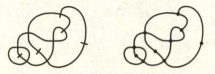

Figure 13.15. The final graph for the game of Brussels sprouts in figure 13.14.

We saw that a game of sprouts must end, but this is not so obvious in Brussels sprouts. Each move destroys two free arms and adds two more, so it seems the game could go on forever. However, every game of Brussels sprouts is destined to end, and herein lies the joke. Regardless of how clever or dim the players, the game will always last for exactly $5n - 2$ turns. In other words, an odd number of starting crosses guarantees the first player to be the winner, and an even number guarantees victor's spoils to the second player.

If we ignore the free arms, then at each turn, the game board is a planar graph (perhaps having several connected components). If we take each cross to be a vertex, then each move adds two edges and one vertex. Unless the new curve joins two connected components, each move also adds a face.

We claim that play will stop when the graph is connected and has exactly $4n$ faces (counting the exterior region as a face). At the start of the game there are n crosses having $4n$ free arms. Because each play wipes out two free arms and adds two more, there are always $4n$ free arms. Each region must have at least one free arm pointing inside it, namely the one coming from its most recently drawn boundary curve. So the graph can have at most $4n$ faces. On the other hand, if there are fewer than $4n$ faces, then some region must contain two free arms, and the game is not yet over.

Suppose the game ends after m turns, leaving a connected graph with V vertices, E edges, and $F = 4n$ faces. (The final configuration in figure 13.15 has $V = 10$, $E = 16$, and $F = 8$.) As we explained earlier, each of the m turns adds two new edges and one new vertex. Because the game began with no edges and n vertices, at the end of the game there are $E = 2m$ edges and $V = n + m$ vertices. Now we use Euler's formula to obtain

$$2 = V - E + F = (n + m) - 2m + 4n.$$

Solving for m gives $m = 5n - 2$.

So, if you feel the need to swindle your unsuspecting friends, ask them to play Brussels sprouts. Give them the choice: they can choose who goes first or the number of starting crosses. Either way, you can guarantee that you are the winner.

CHAPTER 14

IT'S A COLORFUL WORLD

*"Illinois is green, Indiana is pink. You show me any pink
down there, if you can. No sir, its green ... "*
*"Huck Finn, did you reckon the States was the same
color out of doors as on the map?"*
*"Tom Sawyer, what's a map for? Ain't it to learn you
the facts?"*
"Of course."
*"Well, then, how's it going to do that if it tells us lies?
That's what I want to know."*
—Mark Twain, *Tom Sawyer Abroad*[1]

The mathematician Charles Lutwidge Dodgson (1832–1898), better
known as Lewis Carroll, author of *Alice in Wonderland*, invented the
following two-person game. Person A draws a map of a continent with any
number of countries. Then person B colors the map so that neighboring
countries (they have to share a border, touching at a corner does not count)
have different colors. The object of the game is for A to draw a map so
complicated that it forces B to use many colors. B, meanwhile, has to figure
out how to color the map with as few colors as possible.

A simple map resembling a chess board can be colored using only two
colors, but even the most unskilled map-drawer can force three colors. It
is also not difficult to draw a map requiring four colors—simply draw a
country surrounded by exactly three others (such as how Luxembourg is
surrounded by Germany, France, and Belgium). Because all four countries
are mutual neighbors, four colors are necessary (Paraguay and Malawi are
two other examples).

In the introduction we saw a more subtle way to force four colors. Four
colors are necessary to color Nevada, California, Arizona, Utah, Idaho,
and Oregon because Nevada is surrounded by an odd number of states
(figure 14.1). It turns out that Nevada, West Virginia, and Kentucky are
the only such problem states in the United States. By carefully coloring

Figure 14.1. The map of the United States can be colored using four colors.

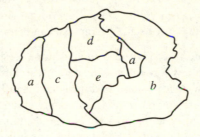

Figure 14.2. Disconnected countries can force five colors.

these states and their neighbors, it is possible to extend the coloring to the entire United States using a fourth color only two times.

Can person A do better than that? Can A force B to use five colors? As we will discuss shortly, it is impossible to find five mutually adjacent countries (we decide in advance that we do not allow disconnected "imperialistic" nations, such as country *a* in figure 14.2). Is that enough to ensure that four colors are sufficient for coloring any map?

According to folklore, mapmakers were the first to notice that four colors sufficed to color any map. There is no evidence to support this assertion. If they were aware of it, they did not publicize it. Kenneth O. May scoured numerous books on cartography and the history of map making and found no mention of the four color theorem.[2] A quick perusal of an atlas shows that most maps are colored using either one color or many colors. There is no indication that mapmakers felt any compulsion to minimize the number of colors used.

As far as anyone can tell, the observation dates back to 1852. Francis Guthrie (1831–1899), a recent graduate in mathematics, recognized that all the counties in England could be colored using only four colors, and he wondered if this was always true. He formulated the conjecture that would become one of the trickiest and most well-known problems in all of mathematics: *the four color conjecture*.

FOUR COLOR CONJECTURE
Every map can be colored with four or fewer colors so that no neighboring nations are the same color.

Francis Guthrie mentioned this observation to his brother Frederick, who in turn shared it with his professor, the respected mathematician Augustus De Morgan (1806–1871). De Morgan was intrigued. On October 23, 1852, he wrote to Sir William Rowan Hamilton (1805–1865):

> A student of mine asked me today to give him a reason for a fact which I did not know was a fact—and do not yet. He says that if a figure be anyhow divided, and the compartments differently colored so that figures with any portion of common boundary line are differently colored—four colors may be wanted, but not more ... Query cannot a necessity for five or more be invented?... What do you say? And has it, if true, been noticed?... The more I think of it the more evident it seems. If you retort with some very simple case which makes me out a stupid animal, I think I must do as the Sphynx did.[3]

Fortunately for De Morgan, he did not have to leap to his death as did the Sphynx after Oedipus answered her riddle correctly. Indeed, Hamilton was not even tempted by the problem. He replied, "I am not likely to attempt your 'quaternion of colors' very soon."[4]

Although De Morgan tried get a few other people to work on the problem, the mathematical community stubbornly refused to embrace it. Nothing appeared in print for almost twenty years. The tipping point for the four color problem was on June 13, 1878, when, at a meeting of the London Mathematical Society, the esteemed mathematician Arthur Cayley asked if the problem had been solved and admitted that he could not do it. The question was then printed and distributed widely in the Society's proceedings.[5]

This problem has been a favorite of mathematical enthusiasts ever since Cayley brought it to the attention of the world. The beauty of the question is that it is so easy to state that a young child could understand it. It is a mathematics problem, to be sure, but it requires no arithmetic, algebra, trignometry, or calculus. Its proof seems tanatalizingly within reach. As renowned geometer H. S. M. Coxeter (1907–2003) wrote:

> Almost every mathematician must have experienced one glorious
> night when he thought he had discovered a proof, only to find
> in the morning that he had fallen into a similar trap.[6]

While at *Scientific American*, Martin Gardner used to receive one lengthy proof of the four color conjecture every few months (all flawed, of course). So in 1975, when he decided to write an April Fools' Day column, he included the four color conjecture. In his article "Six Sensational Discoveries that Somehow or Another have Escaped Public Attention," he reported on six major discoveries of 1974, one of which was a counterexample to the four color conjecture. The caption under the 110-region map said it all: "The four-color-map theorem is exploded."[7] The joke was lost on many readers, however. He received over a thousand letters about this article, including more than a hundred colored copies of the "counterexample," from dupes who did not recognize the hoax.

Although we now attribute the four color problem to Francis Guthrie, many older texts erroneously give credit to the German mathematician August Möbius (1790–1868). Möbius, a descendant of Martin Luther, was a quiet and reserved family man. As an adult he did not travel often, but as a graduate student he attended Leipzig University, the University of Göttingen (he worked with Gauss for two semesters), Halle University, and then back to Leipzig where he completed his doctoral thesis in astronomy. After this period of frequent relocations, he vowed to remain in his beloved Saxony. Despite repeated job offers from other universities, he stayed at Leipzig for the rest of his career.

At Leipzig Möbius worked as an astronomer and oversaw their observatory. He loved mathematics, and it was in mathematics, not astronomy, that he made his most important contributions. He is best known for his work on barycentric calculus, projective and affine geometry, and the foundations of topology. His solitary and careful approach to mathematics produced fine work, but did not make him a gifted lecturer. Because of this, very few fee-paying students attended his classes.

The source of this misattribution is a story told by one of Möbius's students, Richard Baltzer (1818–1887). Baltzer wrote that in 1840 Möbius posed to his class the *problem of the five princes*. He described it

Figure 14.3. August Möbius.

as follows:

> There was once a king in India who had a large kingdom and five sons. In his will he decreed that, after his death, the kingdom should be divided among his sons in such a way that the territory of each should have a common boundary line (not merely a point) with the territories of the remaining four. How was the kingdom divided?[8]

The following day Möbius confessed to his frustrated students that the problem was impossible as stated.

Given what we know, it is easy to deduce the impossibility of a solution. Suppose there was a way to divide the kingdom into five such regions, and in each region was the prince's palace. Then for each pair of brothers it is possible to build a road between their two palaces so that the road does not cross the other brothers' land. However, this creates a planar graph with palaces as vertices and roads as edges. Specifically, it is the complete graph on five vertices, K_5, which we proved is not planar.

In a postscript Baltzer drew the false conclusion that the impossibility of this solution implies that the four color theorem is true. He wrote, "How delighted Möbius would have been to see such a far-reaching application" of this problem.[9] The tenuous link between the five-princes problem and the four color problem is that if such a partitioning of the kingdom existed, then (just like the map in figure 14.2) it would be impossible to color the

Figure 14.4. The adjacency graph for a map.

map with only four colors. However, in reality this eliminates only one of the barriers in the proof of the four color theorem. It may still be possible to produce a complicated map without five mutually neighboring regions that cannot be colored with four colors. According to Martin Gardner, many of the false proofs of the four color theorem that arrived in his mailbox were nothing more than the five-princes problem in disguise.

We should not toss out Möbius's puzzle entirely. The technique of joining palaces by roads is actually a useful one. Just as with the Königsberg bridge problem, the precise geography of the countries is irrelevant; all that is important are the relative locations. It is a topological problem that can be repackaged in terms of graph theory.

The *adjacency graph* of a map has one vertex for each country, and two vertices are joined by an edge whenever the corresponding countries share a border (see figure 14.4). If two countries share more than one border, we still draw only one edge between the corresponding vertices.

Adjacency graphs have some nice properties. It is not difficult to see that the adjacency graph of any map is planar. Simply take the vertices to be the capitals of the countries and the edges to be roads between the capitals that stay inside the two countries. By construction, an adjacency graph will be devoid of loops and parallel edges; such a graph is called *simple*. In short, an adjacency graph for a map is a simple planar graph. Notice that if the map is connected, then the adjacency graph will be, as well.

When we create the adjacency graph for a map, we transform a map-coloring problem into a graph-coloring problem. Instead of coloring the countries of the map, we color vertices of the graph. If a map (or graph) can be colored using *n* colors so that neighboring countries (vertices) have different colors, we will say that it is *n-colored*. We may restate the four

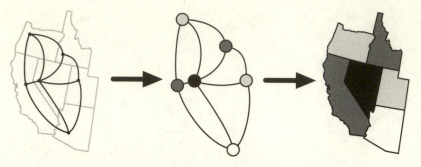

Figure 14.5. Coloring the adjacency graph gives a coloring of the map.

color conjecture as follows.

> **FOUR COLOR CONJECTURE FOR PLANAR GRAPHS**
> Every simple planar graph can be 4-colored.

In figure 14.5 we see the map of Nevada and its neighbors and the associated adjacency graph. We color the graph using four colors, then transfer this coloring to the original map.

In a typical map there may be countries with many neighbors, but this is impossible for all countries. In any map there must be some country that has five or fewer neighbors. We call this important fact the *five neighbors theorem*. The proof of the five neighbors theorem uses Euler's formula and a little counting. In terms of graphs we state it as follows.

> **FIVE NEIGHBORS THEOREM**
> Every simple planar graph has a vertex of degree five or less.

Suppose we have a simple planar graph. Because the graph has no loops or parallel edges, we can add edges to it so that every face is bounded by exactly three edges. We will prove that this (larger) triangulated graph has a vertex of degree 5 or less, and therefore so must the (smaller) original graph. Suppose the triangulated graph has V vertices, E edges, and F faces (counting the exterior region as a face). Each edge borders two faces and each face is bounded by three edges, so $3F = 2E$. By Euler's formula $V - E + F = 2$, or equivalently, $6E - 6F = 6V - 12$. Substituting $4E$ for $6F$ we obtain

$$2E = 6V - 12.$$

Because each edge has two endpoints, the sum of the degrees of all vertices is $2E$. So the average degree of the vertices is

$$\text{average degree} = \frac{2E}{V} = \frac{6V - 12}{V} = 6 - \frac{12}{V} < 6.$$

Of course, because the average degree is less than six, there must be at least one vertex of degree five or less.

To show how the five neighbors theorem is useful in graph-coloring problems, we will prove the six color theorem.

> SIX COLOR THEOREM
> Every map can be colored with six or fewer colors.

For the sake of contradiction, assume that this statement is false. Then there are one or more maps that cannot be six-colored. Survey this collection of nasty maps and find one that has the fewest number of countries. Suppose this map has N countries. Such a smallest counterexample is often called a *minimal criminal*. The benefit of singling out a minimal criminal is that we can say with certainty that any map with $N-1$ or fewer countries can be six-colored.

Consider the adjacency graph G for this minimal criminal. By the five neighbors theorem, there is some vertex v in G with degree five or less. Remove v and all edges incident upon v from G to obtain a new graph H. It is not difficult to see that H is the adjacency graph for a map with $N-1$ countries. Since H has $N-1$ vertices, it can be six-colored. Now, put the vertex and edges back into the graph. Since v is adjacent to at most 5 other vertices, there is at least one unused color remaining to color v. Thus it is possible to six-color G. This contradicts our assumption that G was a minimal criminal, so every map is six-colorable. In figure 14.6 we use this technique to color a graph using red, blue, yellow, green, purple, and orange.

Unfortunately, this same proof does not work when there are only four or five colors available. When it comes time to reinsert the vertex v, there may be no color remaining to color the vertex. We must employ a more subtle trick for these cases.

One such trick was discovered by Alfred Bray Kempe (1849–1922). On July 17, 1879, Kempe, a student of Cayley's, announced that he had a proof of the four color conjecture, and his proof was published later that year.[10]

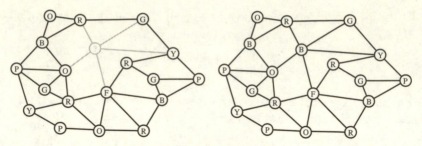

Figure 14.6. Six-coloring a minimal criminal.

Figure 14.7. Alfred Bray Kempe.

Unlike most of the false proofs that followed in the next hundred years, Kempe's proof was very convincing. He introduced clever new techniques that enabled him to color the remaining vertex of the minimal criminal. The mathematical community was thrilled by the proof.

Kempe's proof remained the final word on the four color conjecture for a decade. Unfortunately for Kempe, the case was not yet closed. In 1889 Percy John Heawood (1861–1955) discovered an error in Kempe's argument. It turned out to be a fatal flaw. Heawood produced a map for which Kempe's logic fell apart. In his published note, which appeared in 1890, Heawood wrote:

> The present article does not profess to give a proof of this original Theorem; in fact its aims are so far rather destructive than constructive, for it will be shown that there is a defect in the now apparently recognized proof.[11]

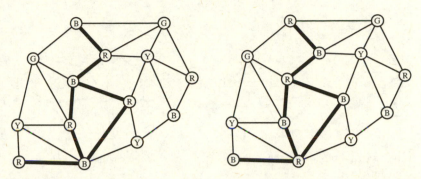

Figure 14.8. Interchanging the colors of a red-blue chain produces another valid coloring.

And so the four color theorem became the four color conjecture again.

Although Kempe's proof was wrong, the techniques he introduced were very important. Heawood acknowledged that Kempe's ideas were sufficient to prove the five color theorem. In fact, they were essential ingredients in the eventual proof of the four color theorem. Although this false proof may have been an embarrassment to Kempe, it did not permanently damage his career. He continued to be an active member of the Royal Society (he had been elected for mathematical work unrelated to the four color theorem) and he was later knighted.

As anyone who has tried to four-color a large map knows, it is easy to color the map for a while but then get stuck so that it is impossible to complete the coloring.* At this point the colorer must back up and re-color parts of the map. The trick that Kempe gave us is an easy way to recolor a map.

Start with any colored (or partially colored) graph. Pick two colors, say red and blue, and a vertex of one of these colors. Follow all the possible paths from this vertex that pass through a blue vertex, a red vertex, a blue vertex, and so on. This collection of red and blue vertices is called a red-blue *chain*, or a *Kempe chain* (see figure 14.8). Notice that a Kempe chain is often not linear; it may have branches or loops. The key observation is that because no vertex adjacent to such a Kempe chain is red or blue, we

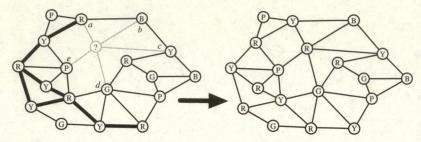

Figure 14.9. Interchange the colors of the red-yellow chain to complete the coloring.

may change every red vertex in the chain to blue and every blue vertex in the chain to red and still have a valid coloring of the graph.

Earlier we proved the six color theorem. This trick of Kempe's enables us to prove the five color theorem.

> FIVE COLOR THEOREM
> Every map can be colored with five or fewer colors.

We begin the proof in exactly the same way that we did for the six color theorem. Suppose we have a minimal criminal—a map with the fewest number countries, N, that cannot be five-colored. Again, by the five neighbors theorem there is a vertex v in the adjacency graph G that has degree five or less. Let H be the graph obtained by removing the vertex v. Because H has $N - 1$ vertices, it can be five-colored. Consider the vertices adjacent to v. If there are 4 or fewer colors used to color these vertices (for instance, if the degree of v is 4 or less), then we may complete the coloring by choosing an unused color to color v. If the vertices adjacent to v use all five colors then our solution is not so simple.

Suppose the vertices adjacent to v are named a, b, c, d, and e (labeled clockwise), and they are colored red, blue, yellow, green, and purple, respectively. Consider the red vertex a and the red-yellow chain containing it. There are two cases to examine. First, as in figure 14.9, suppose that vertex c is not in this red-yellow chain. Then we can interchange the red and yellow vertices in the chain without changing the color of vertex c. In particular, we can then color v red and obtain a five-coloring of G.

On the other hand, suppose c is in this red-yellow chain (as in figure 14.10). Then changing the colors in the chain would also change the color of c and not free up a color for v. This would not help at all.

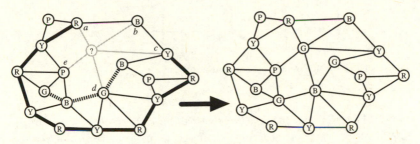

Figure 14.10. Interchange the colors of the blue-green chain to complete the coloring.

However, because the graph is planar, the blue-green chain containing vertex d cannot contain the vertex b. Thus, interchanging the colors in this blue-green chain allows v to be colored green, and we obtain a five-coloring of G.

Kempe's faulty proof of the four color theorem was similar to this proof of the five color theorem. However, the argument was necessarily more subtle. It was possible to have a vertex v of degree 5 surrounded by vertices of four different colors. He had to recolor one or two Kempe chains to decrease that number to three so that he could color v. Although his method looked correct, he missed one case where recoloring two Kempe chains could yield an inadmissible coloring.

The popularity of this fascinating problem continued to enthrall mathematicians and novices alike. Notable mathematicians such as George D. Birkhoff (1884–1944), Hassler Whitney (1907–1989), Henri Lebesgue (1875–1941), and Oswald Veblen (1880–1960) threw their hats in the ring. Despite their long lists of accomplishments, these giants were unable to crack the difficult problem. Some well-respected mathematicians, such as H. S. M. Coxeter, even expressed doubts that the conjecture was true.

As the twentieth century progressed, attention turned to *unavoidable sets* and *reducible configurations*. An unavoidable set is a collection of configurations, at least one of which must be present in every adjacency graph. For example, the five neighbors theorem gives us the simplest unavoidable set, shown in figure 14.11—there must be a vertex of degree less than six.

On the other hand, a reducible configuration is a collection of vertices and edges that cannot appear in a minimal criminal. By using the method of Kempe chains it is easy to show that the first four configurations in

Figure 14.11. An unavoidable set of configurations.

figure 14.11 are reducible. We can remove the vertex, color the rest of the graph, recolor once using Kempe chains, if necessary, then color the last vertex. The fifth one is the problem.

Thus, the goal became finding an unavoidable set of reducible configurations. Doing so would prove the four color theorem because it would be a collection of configurations that could not appear in a minimal criminal, yet must appear in every adjacency graph. This would produce a contradiction to the existence of a minimal criminal.

On July 22, 1976, almost a century after Kempe's failed proof, two researchers from the University of Illinois, Kenneth Appel (b. 1932) and Wolfgang Haken (b. 1928), announced that they had found an unavoidable set containing 1,936 reducible configurations. By the time their two papers appeared the following year they were able to simplify their work and remove redundancies, reducing the number to 1,482.[13] (They also added a third author to one of the papers, John Koch, for his help with computations.) The four color theorem had fallen, at last!

> FOUR COLOR THEOREM
> Every map can be colored with four or fewer colors.

At the end of the summer of 1976 Haken presented their work to the attendees of a joint meeting of the American Mathematical Society and the Mathematical Association of America. At the end of the lecture the audience did not burst into applause, whoop for joy, or enthusiastically pat Haken on the back. Instead, they responded with polite applause. For the roomful of theoretical mathematicians, the much-anticipated conclusion to one of the most interesting stories in mathematics was extremely anticlimactic.

The reason for the cool response was that after Appel and Haken compiled the graph configurations, which filled seven hundred hand-written pages, they plugged them into a computer and let it check the many thousands of special cases. The work of the computer was not even remotely checkable by hand. The computations took six months, used over

Figure 14.12. Kenneth Appel and Wolfgang Haken.

one thousand hours of computer time, and produced a stack of printouts four feet tall. Although most people came to believe that their proof was correct, most pure mathematicians found the proof inelegant, unsatisfying, and unsporting. It was as if Evel Knievel boasted that he could cross the Grand Canyon on his motorcycle, only to build a bridge and use it to make the crossing. Perhaps it is how mountain climbing purists feel about the use of bottled oxygen in high-altitude climbing.

Scientists and engineers have used computers to solve countless problems, but mathematicians have not. Computers are good at making speedy calculations, but not at the kind of precise and subtle arguments that are required in mathematical proofs. Like writing, philosophy, and art, mathematics has always been a human endeavor, one that cannot be automated. Perhaps some day someone will create a black box that proves theorems. We put a statement in, and the black box says "true" or "false." (There are some early attempts at this.) Some would say that this would take the fun out of mathematics and make it less beautiful.

The four color proof was the first high-profile computer-assisted proof. These show no inclination of going away. Another controversial example is Thomas C. Hales's 1998 proof of the Kepler conjecture.[14] Hale proved that Kepler was correct when he claimed that the most efficient way to pack spheres in a box is by staggering them in a crystalline pattern, just as grocers do with oranges or artillerymen did with cannon balls. Although the result appeared in the prestigious *Annals of Mathematics*, it took years for the journal to agree to publish it (it came to press in 2005), and even then the editors said that they did not and could not verify the thousands of lines of computer code.

In the years since Appel and Haken unveiled their controversial proof, it has been independently verified. Other mathematicians have found smaller unavoidable sets of reducible configurations and have discovered more efficient ways of proving the theorem, but thus far every proof has required computer verification.

Paul Erdős (1913–1996), the famously eccentric Hungarian mathematician, used to speak of "The Book"—the imaginary tome that contains the most beautiful and elegant proofs of mathematical theorems. Today the door to the four color theorem is almost closed, but we are still waiting for an old fashioned, pencil-and-paper verification—we have not yet seen the proof from The Book.

CHAPTER 15

NEW PROBLEMS AND NEW PROOFS

The first important notions in topology were acquired in the course of the study of polyhedra.
—Henri Lebesgue[1]

Suppose you were asked: what trees change color and lose their leaves in the autumn? If you replied, "Maple trees do," then you would have given a correct answer. However, anyone who has driven through the Pennsylvania countryside in October knows that there are also radiantly colored oak, birch, and beech trees standing amid beds of fallen leaves. So, although the answer was correct, it did not give a complete account of all such trees. Could you say that all trees change color in the autumn? No. Pine, fir, spruce, and cedar trees have no leaves to lose. In order to make a general, but true, statement, one must look closely at numerous trees. A more complete answer would be: deciduous trees change color and lose their leaves in the autumn.

Convex polyhedra satisfy $V - E + F = 2$. This is a true statement. We know this from the proofs of Euler, Legendre, Cauchy, and others. However, we know that this is not the best we can do. As Poinsot pointed out, Euler's formula holds for more polyhedra than just those that are convex—star-convex polyhedra, for instance. The mathematician D. M. Y. Sommerville (1879–1934) wrote, "Convexity is to a certain extent accidental, and a convex polyhedron might be transformed, for example, by a dent or by pushing in one or more of the vertices, into a nonconvex polyhedron with the same configurational numbers."[2] So, it is misleading and needlessly simplistic to say that convex polyhedra are the only ones that satisfy Euler's formula. Ernest de Jonquières believed that "in invoking Legendre, and like high authorities, one only fosters a widely spread prejudice that has captured even some of the best intellects: that the domain of validity of the Euler theorem consists only of convex polyhedra."[3]

Can we go so far as to say that Euler's formula holds for all polyhedra? No, just as there are trees that do not change color in the autumn, there are polyhedra that do not satisfy Euler's formula. We would like to determine exactly what properties a polyhedron must possess to satisfy Euler's formula. The mineralogist Johann Friedrich Christian Hessel (1796–1872), whom we will meet shortly, called such polyhedra *Eulerian*.

As we discussed in chapter 2, mathematicians worked with polyhedra for many centuries without having a proper definition. They were safe so long as they assumed (almost always implicitly) convexity, but once they tried to claim that something was true for all polyhedra, they generally ran into trouble. The need for a rigorous definition of polyhedra came to a head in the early nineteenth century.

The first person to look carefully at which polyhedra satisfy the polyhedron formula was Simon-Antoine-Jean Lhuilier (1750–1840). Perhaps Lhuilier was destined to work on Euler's formula. Like Euler, Lhuilier was Swiss, and the year of his birth was the year that Euler discovered the polyhedron formula. Most amusingly, the literal translation of *l'huilier* is "the oilcan" or "the one who oils," so Lhuilier may be called "The Oiler."

Like Euler, Lhuilier was tempted away from the clergy by mathematics. When Lhuilier was young, one of his relatives offered to leave him part of his fortune if he pursued a career in the church. Instead of taking this generous offer, Lhuilier decided to become a mathematician.

Lhuilier spent the early part of his mathematical career in Warsaw as a tutor for the son of Prince Adam Czartoryski. Afterward he returned to Switzerland where he took a position at the Geneva Academy, eventually rising to the level of rector. During his long life he contributed to geometry, algebra, and probability, and from this work he received international recognition. He also wrote popular textbooks that were used for many years in Poland. About his personality one biographer wrote, "Whereas the Poles found Lhuilier distinctly puritanical, his fellow citizens of Geneva reproached him for his lack of austerity and his whimsicality, although the latter quality never went beyond putting geometric theorems into verse and writing ballads on the number three and on the square root of minus one."[4]

In 1813 Lhuilier made an important contribution to the theory of polyhedra and to understanding Euler's formula. In his paper he presented three classes of polyhedra that failed to satisfy the polyhedron formula. He called them "exceptions."

Lhuilier's paper was published in the new, privately-established journal *Annales de mathématiques pures et appliquées*. This journal, the first one

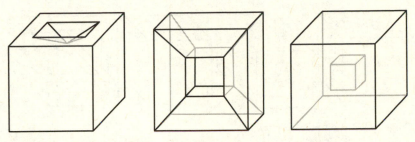

Figure 15.1. Lhuilier's exceptions: annular faces, tunnels, and cavities.

dedicated solely to mathematics, was founded and edited by the French artillery officer and accomplished geometer Joseph Diaz Gergonne. As the mathematician Jean-Claude Pont wrote, Gergonne "had the detestable habit to publish, of work that one submitted to him, only the parts that interested him."[5] Not only did Gergonne edit Lhuilier's work substantially, he repeatedly interjected his own comments into the text of Lhuilier's article—even the assertion that he knew two of the three exceptions before reading Lhuilier's paper!

The first class of exceptions presented by Lhuilier consisted of polyhedra with annular, or ring-shaped, faces. For instance, in figure 15.1 an indentation in the middle of one face of a cube produces a face in the shape of a square washer. This polyhedron has 13 vertices, 20 edges, and 10 faces (5 square faces, 4 triangular faces, and the single annular face). In this case Euler's formula does not hold because $13 - 20 + 10 = 3$. Lhuilier did not call these faces annular or ring-shaped. He remarked that the face has an "inner polygon."

The second class of exceptions presented by Lhuilier is polyhedra with one or more "tunnels" bored through the center. In figure 15.1 we see a polyhedron in the shape of a polyhedral doughnut. In this example the polyhedron has 16 vertices, 32 edges, and 16 faces, thus $16 - 32 + 16 = 0$.

The inspiration for the third class of exceptions came from one of Lhuilier's friend's mineral collection. In one of the specimens Lhuilier saw a colored crystal suspended inside a clear crystal. (Later in 1832, Hessel was also inspired by such a crystal—in his case a lead sulphide cube within a calcium chloride crystal.) Lhuilier imagined a polyhedron with a polyhedron-shaped cavity in the interior. Of course this exception only makes sense if a polyhedron is assumed to be solid and not hollow. A cube with a cube-shaped cavity is shown in figure 15.1. This polyhedron has 16 vertices, 24 edges, and 12 faces, and $16 - 24 + 12 = 4$.

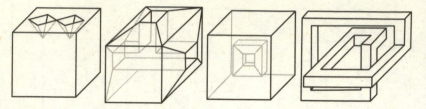

Figure 15.2. Complicated polyhedra.

Lhuilier (and Gergonne) believed that this collection encompassed all of the possible exceptions to Euler's formula. Lhuilier wrote, "One will easily be convinced that Euler's theorem is true in general for all polyehdra, whether they are convex or not, except for those instances that will be specified."[6]

Then, rather than ignoring the exceptions, Lhuilier devised a modification of Euler's formula that took into account the features of his exceptional polyhedra. He asserted that a polyhedron with T tunnels, C cavities, and P inner polygons satisfies

$$V - E + F = 2 - 2T + P + 2C.$$

A quick count shows that this formula is indeed valid for the three polyhedra in figure 15.1.

It turns out, however, that Lhuilier's three exceptions do not include all of the possible exceptions to Euler's formula, and that his insightful formula does not apply to all "exotic" polyhedra. For instance, the four polyhedra in figure 15.2 do not fit into Lhuilier's three categories and it is not clear how to apply his formula. The first polyhedron has a face with two inner polygons that share a common vertex; the second has a tunnel with a branch in it; the third has a cavity in the shape of a torus; and the fourth is shaped like a torus, but the tunnel is not obvious.

Again, we return to the problem of defining polyhedron—it is impossible to classify Eulerian polyhedra without first having an accurate definition of polyhedron. Nevertheless, Lhuilier's classification of these exceptions was extremely important, and a suitably modified version of his formula would eventually be correct. In fact, according to Lakatos, this modified version of Euler's formula, or one like it, was rediscovered a dozen times in the eighty years following Lhuilier's discovery.

Johann Hessel was originally trained in medicine, but he changed his vocation after the noted mineralogist K. C. von Leonard urged him to

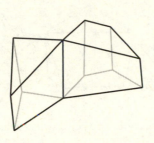 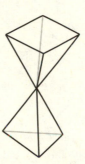

Figure 15.3. Hessel's exceptions to the polyhedron formula.

become a mineralogist. Eventually Hessel became a professor of mineralogy and mining technology in Marburg, Germany. He made contributions to many areas of science, but he is most well known for his mathematical investigation of symmetry classes of minerals.

In his paper of 1832 Hessel presented five exceptions to the polyhedron formula.[7] When he wrote and submitted his paper, he was unaware of Lhuilier's work from two decades earlier. Soon Hessel learned of Lhuilier's work and learned that three of his five exceptions coincided with Lhuilier's. Hessel believed that many people were unaware of these important exceptions, so he decided not to withdraw the publication. Hessel's two new exceptions are shown in figure 15.3. One is a polyhedron formed from two polyhedra joined at an edge, and the other is a polyhedron formed from two polyhedra joined at a vertex. It is debatable whether these figures should be classified as polyhedra, but they surely fail to satisfy the polyhedron formula. The first has 12 vertices, 20 edges, and 11 faces ($12 - 20 + 11 = 3$) and the second has 8 vertices, 14 edges, and 9 faces ($8 - 14 + 9 = 3$).

Louis Poinsot found two more exceptions in 1810.[8] In the same paper that contained the refinement of Legendre's proof, Poinsot presented the four star polyhedra shown in figure 15.4. As we have seen, mathematics is often discovered, forgotten, and then rediscovered. Two of Poinsot's four star polyhedra, the great and small stellated dodecahedra, were described by Kepler (see figure 6.6), and prior to that they appeared in the artwork of Jamnitzer and Uccello (figure 6.3). Poinsot was the first to present the other two star polyhedra, the great dodecahedron and the great icosahedron, in a mathematical context, although the former is also seen in the drawings of Jamnitzer (figure 6.3). These four polyhedra are now referred to as *Kepler-Poinsot polyhedra*.

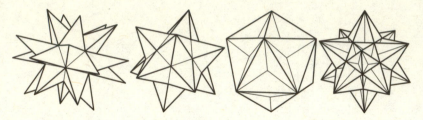

Figure 15.4. The Kepler-Poinsot polyhedra: great and small stellated dodecahedra, the great dodecahedron, and the great icosahedron.

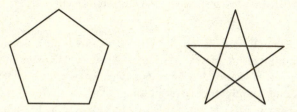

Figure 15.5. A regular pentagon and a regular self-intersecting pentagon, the pentagram.

The obvious way to view these is as nonconvex polyhedra formed from triangular faces. As we have already pointed out, they are star convex and thus, by Legendre's proof, satisfy the polyhedron formula. However, neither Kepler nor Poinsot perceived them in this way. They saw these exotic polyhedra as new regular polyhedra.

In order to understand their point of view, we need to return to polygons in the plane. Earlier we asserted that there is only one regular n-sided polygon for each $n > 2$. For instance, the regular pentagon is shown in figure 15.5. However, if we were to loosen our requirements and allow the sides of the polygon to intersect, then we would find another regular pentagon—the pentagram of the Pythagoreans. After all, drawing a pentagram requires only five strokes of the pencil. We think of the pentagram as having five vertices and five sides joining these vertices. The edges happen to intersect two other sides, but we ignore these intersections and do not count them as vertices. The pentagram is formed by five sides of equal length and the angles between them are equal. It is very reasonable to call this polygon regular.

Kepler and Poinsot viewed their star polyhedra in this same way. Instead of forming the great dodecahedron out of triangles, we construct it from twelve self-intersecting pentagonal faces (see figure 15.6). That is, take all

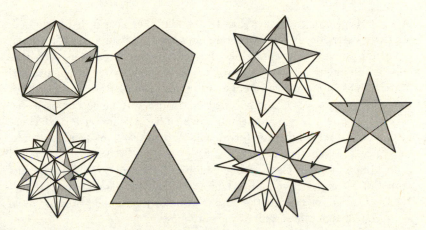

Figure 15.6. Regular polyhedra with self-intersecting faces.

coplanar faces and join them to make a single face. In this way, the great
dodecahedron is constructed from congruent regular pentagons and the
same number of faces meet at each vertex. If we are willing to drop the
convexity requirement, then the great dodecahedron can be considered a
regular polyhedron, just like the Platonic solids. Likewise, the other three
Kepler-Poinsot solids also have this redefined trait of regularity—the great
and small stellated dodecahedra have pentagrams as faces and the great
icosahedron has equilateral triangles as faces.

It turns out that, just like Theaetetus proved that there are only
five regular polyhedra, in 1811 Cauchy proved that only four polyhedra
satisfy this new, relaxed definition of regularity—the four Kepler-Poinsot
polyhedra.[9]

Although these are not everyday polyhedra, we can still compute
$V - E + F$. Sure enough, the great icosahedron ($V = 12$, $E = 30$, $F = 20$)
and the great stellated dodecahedron ($V = 20$, $E = 30$, $F = 12$) satisfy
Euler's formula. However, the other two do not—they are two more exceptions to Euler's formula. Indeed, when we view the great dodecahedron as a
polyhedron with twelve pentagonal faces, it does not satisfy the polyhedron
formula. It has 12 vertices and 30 edges, thus $12 - 30 + 12 = -6$. The small
stellated dodecahedron also has 12 vertices, 30 edges, and 12 faces, so again
the alternating sum yields -6.

The first half of the nineteenth century saw many exceptions to Euler's formula, but it also saw many new proofs. By 1811 there were demonstrations

by Euler, Legendre, and Cauchy. In 1813, in Lhuilier's paper that contained his three exceptions,[10] he gave a new proof that the polyhedron formula holds for convex polyhedra. Like Euler did, Lhuilier decomposed the polyhedron into pyramids. To do so, he placed a new vertex in the interior of the polyhedron and created edges and faces running from this vertex to the vertices and edges of the polyhedron. In this way he decomposed the polyhedron into many pyramids, all of which had a common vertex. Then he proved that the polyhedron formula holds for any pyramid and for solids constructed from pyramids in this way.

In Lhuilier's paper Gergonne gave a proof for convex polyhedra (this same proof was rediscovered fourteen years later by Jakob Steiner [1796–1863]).[11] Gergonne projects the polyhedron down onto the plane and uses an argument about the angles of the polygons.

One of the most clever proofs of the polyhedron formula is due to Karl Georg Christian von Staudt (1798–1867) in 1847. His proof had the special advantage that it applied to a broad class of nonconvex polyhedra.

Staudt was born into a noble family in Rothenburg, Germany. At the age of twenty he enrolled at the University of Göttingen to study astronomy and mathematics with Gauss. His PhD work in astronomy so impressed Gauss that he helped find Staudt a lecturing position at the University of Würzburg while Staudt was working as a secondary-school teacher. In 1835 Staudt became a full professor at the University of Erlangen, where he was the preeminent mathematician. Staudt was not a prolific mathematician, but in 1847 he wrote an influential book on projective geometry called *Geometrie der Lage*, which was later followed by three long supplements. It is for this work that he is best remembered.

Convexity is a sufficient condition for Euler's formula, but as Poinsot pointed out, it is not necessary. In *Geometrie der Lage* Staudt finally gave a very general set of criteria that describe Eulerian polyhedra.[12] Staudt implicitly assumed that his polyhedra were hollow shells, not solid. In addition, he made the following assumptions about his polyhedra:

1. It is possible to get from any vertex to any other vertex by a path of edges.
2. Any path of edges that begins and ends at the same vertex without visiting any vertex twice (recall that such a path is called a circuit) divides the polyhedron into two pieces.

Staudt's insightful criteria include many polyhedra that are not convex. For instance, in figure 15.7 we see two very complicated polyhedra. The first, it turns out, satisfies all of Staudt's criteria (indeed $V = 48$, $E = 72$,

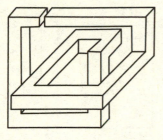 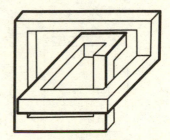

Figure 15.7. Two complicated polyhedra.

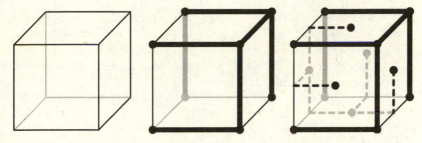

Figure 15.8. The two trees in Staudt's proof.

and $F = 26$, and $48 - 72 + 26 = 2$). The second does not. Cutting along the darkened path of edges does not disconnect the polyhedron (for this polyhedron $V = 40$, $E = 60$, and $F = 20$, and $40 - 60 + 20 = 0$).

Staudt then gave a beautiful argument proving that any polyhedron meeting these criteria must satisfy the polyhedron formula. We now give a brief sketch of this proof.

Color a vertex of the polyhedron red. Beginning at this vertex, color one of the adjacent edges and its other vertex red (the process is illustrated for a cube in figure 15.8 with the thick, solid lines representing red edges). Then pick one of the two red vertices and color an adjacent edge red, including the opposite vertex. Continue coloring edges and vertices in this way obeying one important condition: never create a red circuit. Eventually this process will end. For any polyhedron satisfying Staudt's conditions this will occur precisely when all V vertices have been colored red. Because this collection of red edges has no circuit, it is a tree, and as we saw earlier (figure 13.3 and accompanying text), there must be $V - 1$ red edges.

Now, place a blue vertex inside of each face. Draw a blue edge from one blue vertex to an adjacent blue vertex whenever they are not separated by a red edge (blue edges are dashed lines in figure 15.8). Again, for a

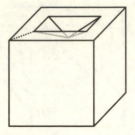 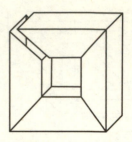

Figure 15.9. Altering an annular face and cutting a torus.

polyhedron satisfying Staudt's criteria, the resulting blue graph will be a tree. This blue tree has F vertices, so it has $F-1$ edges. The crucial observation is that every edge in the original polyhedron is either red or is crossed by a blue edge. Thus the number of edges is the sum of the number of red edges and the number blue edges:

$$E = (V-1) + (F-1),$$

or rearranging terms, $V - E + F = 2$.

We should take a moment and revisit the three exceptions of Lhuilier (shown in figure 15.1) and convince ourselves that they are ruled out by Staudt's definition of polyhedron. Lhuilier's first polyhedron has an annular face. Because it is impossible to get from the outside edges of the annulus to the inside edges, condition number 1 fails. Notice that it is possible to change this polyhedron so that it complies with Staudt's definition. To do so, we simply add an artificial edge that connects the inside of the annular face to the outside (as in figure 15.9).

The second polyhedron has a tunnel through its center. This polyhedron fails condition number 2 because, as we see in figure 15.9, it is possible to cut a circuit of edges that does not disconnect the polyhedron. In 1879 R. Hoppe remarked, "Let the polyhedron be made of some stuff that is easy to cut like soft clay, let a thread be pulled through the tunnel and then through the clay. It will not fall apart."[13] Recall that Lhuilier did not give a proper definition of tunnel. Hoppe used the ideas in Staudt's paper to help rectify this situation. He defined a tunnel in terms of the number of cuts needed to disconnect the surface. We will return to this idea again in chapter 17.

Finally, we can dismiss Lhuilier's third exception easily. This exception, a polyhedron with a polyhedral cavity, makes sense only for solid polyhedra, and Staudt assumed his polyhedra were hollow. Even if he did allow solid polyhedra, Staudt's condition number 1 fails since there are no edges from the inner vertices to the outer vertices. Although Hessel's exceptions satisfy Staudt's two conditions, he, like most mathematicians, did not consider them polyhedra.

Intuitively, the polyhedra that satisfy Staudt's criteria are those that are "sphere-like" and have faces with a single polygonal boundary. The polyhedra need not be convex, but they cannot have any tunnels. If they were made of rubber and were inflated, they would inflate to a spherical balloon.

This fruitful dialogue about Eulerian and non-Eulerian polyhedra in the first half of the nineteenth set the stage for the field that would become topology. These ideas were explored further by others, culminating in Poincaré's marvelous generalization of Euler's formula at the end of the nineteenth century. We will discuss this development in chapters 17, 22, and 23.

CHAPTER 16

RUBBER SHEETS, HOLLOW DOUGHNUTS, AND CRAZY BOTTLES

A mathematician named Klein
Thought the Möbius band was divine.
Said he: "If you glue
The edges of two,
You'll get a weird bottle like mine."
—Anonymous

By the middle of the nineteenth century mathematicians had a much better understanding of how Euler's formula applied to polyhedra. It was during this time that they began to ask whether it applied to other objects. What if the figure was not a polyhedron made of flat faces, but instead was a curved surface like a sphere or a torus? If so, what must the partitions look like? Recall that in 1794 Legendre used a partition of a sphere by geodesic polygons to prove Euler's formula, and Cayley showed that when we apply Euler's formula to graphs, the edges need not be straight.

These discussions illustrate the ongoing transition from a geometric to a topological way of thinking about shapes. The popular press often uses the term "rubber-sheet geometry" to describe the field of topology to a public that is likely to be unfamiliar with the term. Although literal-minded mathematicians take exception to this extreme oversimplification, it is a reasonable way to describe the difference between topology and geometry. In geometry it is crucial that the objects of study be rigid. Measurements of angles and lengths, proofs of congruencies of figures, and computations of areas and volumes all rely on precise and unmoving geometric structure.

As we saw earlier, in some cases the rigid, unbending features of geometric figures are not needed, and worse, they often obscure the underlying mathematics. In Euler's investigation of the bridges of Königsberg, he discovered that it was the general arrangement of features that was important, not their exact locations. This observation led to the creation of graph theory, one of the earliest incarnations of topology. Later we saw hints that the alternating sum $V - E + F$ depended only on the general shape—the topology—of an object, and not on the number of faces or their configuration. We observed that for any sphere-shaped polyhedron, $V - E + F = 2$, for a polyhedron with g "tunnels," $V - E + F = 2 - 2g$, and for any connected planar graph, $V - E + F = 1$.

So it is not difficult to imagine that Euler's formula might apply to shapes other than polyhedra. Begin with a rubber polyhedron that satisfies $V - E + F = 2$. Can we alter this shape so that $V - E + F \neq 2$? Not easily. If we were to inflate it like a balloon so that all the faces and edges became curved, the alternating sum would not change. If we were to squeeze it, twist it, or pull it, the relationships among the numbers of vertices, edges, and faces would remain the same. Only by using a knife to cut a gash in the side of the balloon would the alternating sum change (it would create at least one new edge). In the next chapter we will discuss in more detail what it means for two shapes to be topologically "the same," and we will investigate how Euler's formula applies to various topological shapes.

The mathematical term "topology" dates back to 1847 (it had a botanical meaning earlier than that). It first appeared in German in the title of Listing's book *Vorstudien zur Topologie*,[1] although he had already been using the term in correspondence for ten years. The first appearance in English was in Peter Guthrie Tait's (1831–1901) eulogy of Listing in 1883. He wrote, "The term *Topology* was introduced by Listing to distinguish what may be called qualitative geometry from the ordinary geometry in which quantitative relations chiefly are treated."[2]

The term "topology" did not catch on immediately. Influential mathematicians such as Henri Poincaré and Oswald Veblen continued to use the French term *analysis situs*. The great early-twentieth-century topologist Solomon Lefschetz (1884–1972) was not so enamored with this term. He referred to *analysis situs* as "a beautiful but awkward term."[3]

Lefschetz's route to greatness was a curious one. He was born to Turkish parents in Russia in 1884, was raised and schooled in France, emigrated to America, and took a job as an engineer in Philadelphia. At the age of twenty-six, not long after losing his hands and forearms in a work-related accident, he decided to pursue a career in mathematics. He completed his PhD at Clark University in a single year, and taught briefly in Nebraska before taking a position at the University of Kansas at Lawrence. Then,

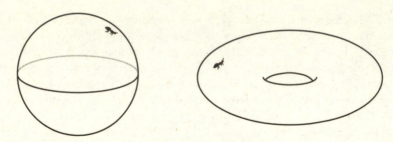

Figure 16.1. Ants on the surface of a sphere and a torus.

at forty, after more than a decade of important work, he was hired by Princeton University. He received numerous honors in his long and distinguished career, including the National Medal of Science.

According to Albert Tucker (1905–1995), one of Lefschetz's students, it was Lefschetz who popularized the use of the term "topology." He titled his influential 1930 book for the American Mathematical Society *Topology*. According to Tucker:

> Lefschetz wanted a distinctive title and also, as he would say, a
> snappy title, so he decided to borrow the word *Topologie* from
> German. This was odd for Lefschetz since he was French trained and
> *analysis situs* was Poincaré's term; but once he decided on it, he
> conducted a campaign to get everyone to use it. His campaign
> succeeded quickly, mainly I think because of the derivative words:
> topologist, topologize, topological. That doesn't go so well with
> *analysis situs*![4]

We begin our investigation of topology by looking at *surfaces*. Examples of surfaces are a 2-dimensional plane, a sphere, a torus, a disk, and a cylinder. A surface is any object that looks locally like a plane. If an ant were to sit on a large surface, it would think that it was sitting on a 2-dimensional expanse. This is not out of our realm of experience—the earth is a spherical globe, but it is so large that to its inhabitants it is indistinguishable from a flat plane. A clever ant may be able to discover that its surface is not flat by venturing out and exploring the surface (just as Columbus tried to do when he sailed west toward "the Indies"), but standing still, it would have no idea.

It is important to be aware of the difference between *intrinsic* and *extrinsic* dimension. As the ant on a surface will tell you, it is locally

2-dimensional—the intrinsic dimension of a surface is two. However, for us to build a physical copy of this surface, the surface must live somewhere, and the dimension of this enveloping space is the extrinsic dimension. The sphere and the torus have an intrinsic dimension of two, but they must live in 3-dimensional space, so their extrinsic dimension is three. Shortly we will encounter bizarre surfaces that cannot be constructed in 3-dimensional space. Their extrinsic dimension is four. From a topological point of view, the intrinsic dimension of a surface is the most important; that is why we say that surfaces are 2-dimensional.

Surfaces are characterized by local simplicity and global complexity. In other words, up close, they are all identical. They all look like the Euclidean plane. However, globally they can differ substantially. They can loop back upon themselves, they can have holes, they can be twisted or knotted, and so forth.

A sphere and a torus are examples of *closed surfaces*. They have no punctures, they do not run off to infinity, and they do not have any sharp boundaries. Sometimes we want to consider surfaces that are not closed. A disk and a cylinder are examples of *surfaces with boundary*. A surface with boundary is still locally 2-dimensional, except that it may have one or more 1-dimensional boundary curves. Some flat-earthers believed that the earth had a boundary. On such a planet the unlucky Columbus would not reach the Indies, but would instead sail off the edge of the ocean.

For simplicity, when we use the term "surface," we will mean *compact* surface. "Compact" is a technical term that means the surface is bounded and contains all of its boundary curves. In other words, we will not consider unbounded surfaces such as the 2-dimensional plane or a piece of cylindrical tubing that runs infinitely far in both directions. When we say that the surface must contain all of its boundary curves, we mean to exclude surfaces such as the open unit disk $(x^2 + y^2 < 1)$. The open unit disk is the set of all points strictly less than one unit away from the origin; it is the unit disk $(x^2 + y^2 \leq 1)$ with the boundary circle removed. A good analogy is the frayed pant legs after the cuffs are removed—we need those cuffs.

In 1882 Felix Klein (1849–1925) devised an ingenious way of constructing surfaces.[5] He began with a polygon (imagine that it is made of a very pliable rubber material). He created a surface by gluing sides of the polygon together in pairs. For example, if we begin with a square, roll it up, and glue together the two opposite sides, we obtain a cylinder (see figure 16.2). Notice that if instead of rolling the square into a cylinder, we were to keep

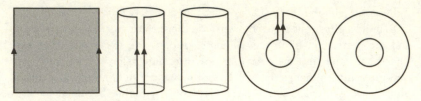

Figure 16.2. A cylinder or annulus.

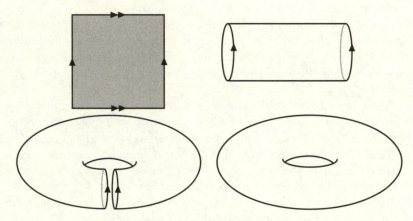

Figure 16.3. Making a torus from a square.

the entire figure in the plane and distort it until the opposite edges meet (we need it to be made of a very soft rubber!), it would form a washer-shaped *annulus*. To a topologist a cylinder and an annulus are indistinguishable.

To make it clear which sides will be glued and with which orientation, it is common to decorate them with arrows. There are two different ways to glue together a pair of sides—with or without a twist. So we use the arrows to show the proper alignment. When we need to glue together more than one pair of sides, we use different numbers of arrows or arrows of different shapes to indicate which pairs attach to which. In figure 16.3 we glue both pairs of opposite sides of a square together. We illustrate this by putting a single arrow on one pair of sides and double arrows on the other pair. First we glue one pair of sides to obtain a cylinder. Then, since the two end circles have compatible orientations, we join these together to obtain a torus.

Some old arcade-style video games, such as Asteroids, employed this representation of a torus. When the space ship flew off the side of the rectangular screen, it would suddenly reappear on the opposite side

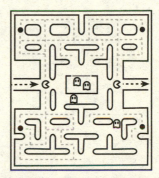

Figure 16.4. Arcade games played on a torus and a cylinder.

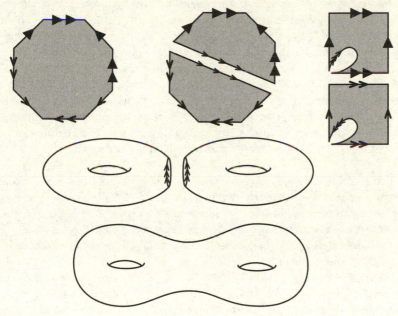

Figure 16.5. A double torus.

(see figure 16.4). If it flew off the top of screen, it would rise from the bottom. Other games had other topological configurations. Pac-Man, for instance, was played on a cylinder.

We do not have to confine ourselves to squares when constructing surfaces. In figure 16.5 we see an octagon with four different pairs of sides identified (they are given by single and double arrows, and one and two

Figure 16.6. A Möbius band.

triangles). To see the shape of the resulting surface, it helps to make an additional cut diagonally through the octagon (we put three arrows on the cuts so that we will be able to glue them back together later). We deform these two pentagons into two squares, each with an indentation. These squares resemble the square in figure 16.3, so after they are glued they each form a torus with a hole in it. Finally, we glue the two tori together along the holes and we obtain a two-holed torus (or double torus).

Klein proved that any surface can be represented as a polygon with sides glued together in pairs, but there may be many polygonal representations of the same surface. Fortunately, each surface has a "nice" polygonal representation such as our examples, and by cutting and regluing, any polygonal representation can be transformed into the nice one.[6]

In each of the examples we have seen thus far, the sides of the polygons have been glued without any twisting. In figure 16.6 we have a square in which a pair of opposite sides are glued together with a twist. Because the square is made of rubber, we can stretch it out, roll it up as if we are making a cylinder, and apply a half-twist before gluing. This shape is the well-known *Möbius band* or *Möbius strip*.

Although the construction of the Möbius band is simple, it has many surprising properties. Unlike the cylinder, the Möbius band has only one side. An ant walking along the centerline of a Möbius band would eventually return to the same location, but would be standing on the other side. Said another way, we can paint a cylinder red on one side and blue on the other, but a Möbius band must be all red or all blue. Also unlike the cylinder, the Möbius band has only one boundary. The ant sees an edge to the left and an edge to the right, but little does it know that they are both the same edge.

The Möbius band is a favorite topological entity for mathematics enthusiasts. It has been reproduced by many sculptors and artists. The

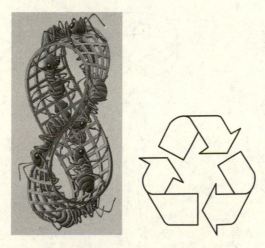

Figure 16.7. Two famous Möbius bands: M. C. Escher's *Möbius Strip II* (1963) and the recycling symbol.

most famous artistic rendering is probably M. C. Escher's (1898–1972) 1963 woodcut featuring (what else!) ants crawling on a Möbius band (figure 16.7). It has appeared in literature—usually science fiction—such as Arthur C. Clarke's 1949 short story "The Wall of Darkness."[7] It formed the basis for Gary Anderson's award-winning design in the 1970 Earth Day contest to create the now-ubiquitous recycling symbol.* It has been used to create conveyor belts and tape loops so that they wear evenly.

The Möbius band is even the basis of a magic trick with the mysterious name of "Afghan bands" that dates back to at least 1882. A circus magician holds up three loops of fabric, which he explains are cloth belts. The problem, he laments, is that he needs belts for two clowns, the fat lady, and the Siamese twins. He takes the first band, rips it down the centerline and produces the belts for the two clowns. He rips the second band in the same way, but instead of two loops, he holds a single loop that has twice the circumference of the original—the belt for the fat lady. Finally, to get the belts for the twins, he rips the third loop and obtains two belts that are linked together. The trick, as we see in figure 16.8, is that the loops have twists in them (zero, one, and two half-twists, respectively). For maximum effect, the fabric or paper should be flexible and be much narrower than

---

* Actually, at some point a variation of Anderson's recycling symbol having three half-twists sprung up. Now it is common to encounter both versions.

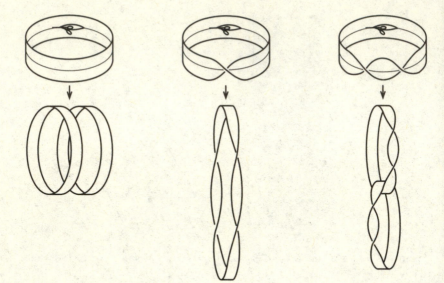

Figure 16.8. Afghan bands.

it is long so that the audience does not notice the twisting. Stephen Barr suggests the following dramatic alteration.[8] Before the show, secretly apply a flammable liquid to the centerline of a twisted loop. In the presence of the audience, tack the edge of the loop to the wall and hold a match to the fabric. After a burst of flame, the loop falls apart into the desired configuration.

The reader is encouraged to set the book aside to try these cutting tricks and other variations (see appendix A for templates). Try giving the bands more than two half-twists. Try cutting a Möbius band along the line 1/3 of the way between the "two" boundaries. The author's personal favorite is a trick due to Stanley Collins.[9] Pass the band through a wedding ring before giving three half-twists and gluing into a band. When this band is cut down the middle it will have a knot in it, and inside the knot is the wedding ring!

The Möbius band is named after Möbius, but it was discovered almost simultaneously by Listing (it was Listing who first observed the mathematics behind what would become the Afghan bands trick). Listing published his description of the Möbius band in 1861,[10] four years earlier than Möbius did.[11] Their correspondences and notes show that the first appearance was in Listing's hand (in July 1858), beating Möbius (September) by a few months.

The reason the Möbius band is not called the Listing band is that Möbius was the first to make mathematical sense of the one-sidedness property

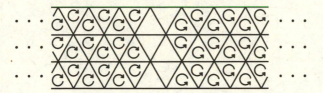

Figure 16.9. A traingulation of a Möbius band cannot be oriented.

Figure 16.10. The Möbius band is not orientable.

of the band. Today we call such surfaces *nonorientable*. There are several ways of describing this phenomenon mathematically. Möbius showed that it is impossible to divide the Möbius band into triangles, then orient the triangles so they match up with their neighbors (see figure 16.9).

Later Klein defined orientability in a different way. Place a small circle on the surface and pick an orientation. This circle is not drawn on one side of the surface, but is part of the surface so it is visible on both sides (on one side it is oriented clockwise, and on the other it is oriented counterclockwise). Imagine that the surface is made of tissue paper and the circle is drawn with a felt-tipped marker which bleeds through to the other side. Klein called such an oriented circle an *indicatrix*. If it is possible to slide this indicatrix around the surface in a such a way that when it returns to its original location, the orientation is opposite, then the surface is nonorientable. In figure 16.10 we show that the Möbius band is nonorientable by sliding the indicatrix around the center circle.

Walther von Dyck (1856–1934), a student of Klein, gave yet another definition. He put a movable coordinate frame, an x- and y-axis, on the surface. If it is possible to move the coordinate frame around the surface in such a way that the axes switch places, then the surface is nonorientable. (One benefit of Dyck's approach is that it easily generalizes to higher-dimensional topological objects.)

It is interesting to note that mathematicians do not use the one-sidedness property to define nonorientability. Although one-sidedness

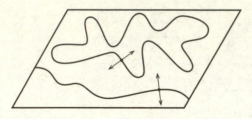

Figure 16.11. A curve in the plane is two sided, but one in 3-dimensional space has no sides.

may seem to be equivalent to nonorientability, Klein and Dyck argued that one-sidedness loses all meaning in higher dimensional spaces, but nonorientability does not. *Sides* only make sense for surfaces living in 3-dimensional space. It is meaningless to refer to the inside or outside of a surface—even a sphere—in 4-dimensional space.

This and other assertions we will make about high-dimensional spaces are difficult to grasp. They require mental gymnastics that human beings are not hard wired to perform. As the mathematician Thomas Banchoff wrote, "All of us are slaves to the prejudices of our own dimension."[12]

To illustrate the perplexing claim that surfaces in 4-dimensional space have no sides, we drop down a dimension from surfaces to curves. As we see in the left-hand image of figure 16.11, for a given point on a planar curve, normal vectors can point in only two possible directions (a vector is *normal* to a curve if it is perpendicular to the line tangent to the curve). So a planar curve has sides, and because it is impossible to move a normal vector around the curve so that it returns pointing in the opposite direction, it is two-sided. If the curve happens to be a *simple closed curve*—a closed loop that does not cross itself—then we call these directions the inside and the outside (actually, the seemingly obvious statement that every simple closed curve has an inside and an outside is a deep theorem known as the *Jordan curve theorem*).

On the other hand, for a curve in 3-dimensional space, there are infinitely many normal directions at each point (as illustrated by the disk of normal vectors in the right-hand image in figure 16.11). So in this case the term "sides" has no meaning.

Similarly, at any point on a surface in 3-dimensional space, there are two normal directions (a normal vector is perpendicular to the plane tangent to the surface). For nonorientable surfaces it is possible to move a normal vector around the surface so that it comes back pointing in the opposite direction, so it is one-sided (see figure 16.12). For orientable surfaces, this

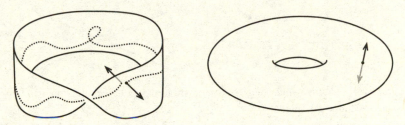

Figure 16.12. In 3-dimensional space the Möbius band is one-sided and the torus is two-sided.

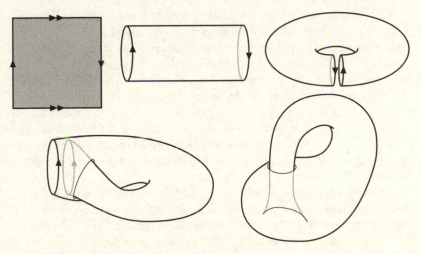

Figure 16.13. A Klein bottle.

is not possible, so they are two-sided. But for a surface in 4-dimensional space, there are infinitely many normal directions at a point, so, as for a curve in 3-dimensional space, "sides" is meaningless.

The Möbius band is not the only nonorientable surface. In 1882 Klein discovered another, this one having no boundary, and it now bears the name the *Klein bottle*.[13] In figure 16.13 we depict it in terms of a square with sides glued. We must glue opposite sides together; the left and right pair is glued with a twist and the top and bottom pair without. To construct the Klein bottle, glue together the two similarly oriented sides to obtain a cylinder. If we were to wrap this cylinder around like a torus, we would find that the ends have opposite orientations. Instead, the cylinder must "pass through" itself and come in from behind so that the end circles line up with the same orientations.

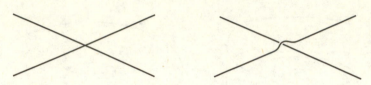

Figure 16.14. With a slight detour in the third dimension, we can allow two lines to pass without crossing.

What do we mean by "pass through"? We do not mean it literally. The Klein bottle is our first example of a surface that cannot be constructed in 3-dimensional space. When we say that the bottle passes through itself, it actually passes by itself in the fourth dimension. To illustrate this baffling concept, we again drop down a dimension. Suppose we wanted to draw two nonparallel lines in the plane that have no point of intersection. Clearly this is impossible, but if we were able to leave the 2-dimensional paper and use the third dimension, then, just before getting to the point of intersection, we could hop over the line (see figure 16.14). Thus, the two lines are essentially planar, but would need a small amount of the third dimension. Using this same technique we can construct the Klein bottle. When it comes to passing the tube through itself, it should make a small hop over itself in the fourth dimension.

Let us return to the square to create one final surface. This last one is the most difficult to visualize. It is formed by gluing together the opposite sides, but both with a twist (figure 16.15). Begin by taking the square piece of rubber and deforming it so that it takes the shape of a bowl. We are careful to keep track of which segments of the boundary should be glued to which other segments. Continue to deform this bowl so that the sides that need to be glued are aligned next to each other with the same orientation. Glue one pair of sides together (in figure 16.15 we glued the sides marked with double arrows). As we can see, we are in a bit of trouble—this gluing forced the remaining pair of sides to be on opposite sides of the new surface. In order to make the final gluing we must make use of the fourth dimension to allow the surface to pass by itself. In figure 16.15 we give two different views of this bizarre nonorientable surface called the *projective plane*.

The projective plane did not first appear in this context—as an object created by the gluing of surfaces. Instead, as the name suggests, it was an object studied in projective geometry, a geometric system in which any two lines, even those that are parallel, meet at a single point. Klein and Ludwig Schläfli (1814–1895) were the first to recognize that the projective plane was nonorientable.

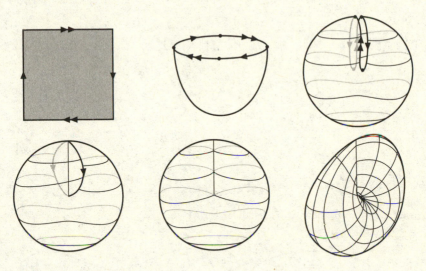

Figure 16.15. The projective plane.

Appendix A contains templates for making the cylinder, the torus, the Möbius band, the Klein bottle, and the projective plane out of paper.

Klein gave one method of creating complicated surfaces from simpler shapes—gluing the sides of polygons together in pairs. We now present another way of constructing complicated surfaces from simpler ones. We start with a sphere and glue onto it cylindrical *handles* to make orientable surfaces and Möbius bands to make nonorientable surfaces.

As we see in figure 16.16, to add a handle to a surface, cut out two disks and glue the ends of a cylinder to the boundaries of the holes. A sphere with one handle is a torus. We build a double torus by adding another handle, and a *g*-holed torus by adding *g* handles to a sphere.

The number of handles on such a surface is intimately related to a topological quantity called the *genus*. The genus of an orientable surface (with or without boundary) is the maximum number of nonintersecting closed curves along which we can cut so that the surface is not disconnected.

To illustrate this concept, consider a sphere. Cutting along any simple closed curve will disconnect the sphere. This is another application of the Jordan curve theorem—just as in the plane, a simple closed curve will divide the sphere into two regions. So the genus is 0. On the other hand, it is possible to cut a loop on the surface of a torus so that it remains connected

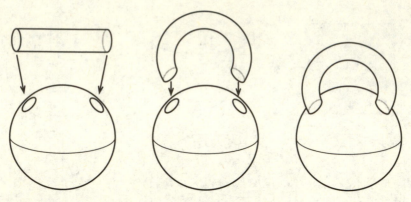

Figure 16.16. A sphere with a handle (a torus).

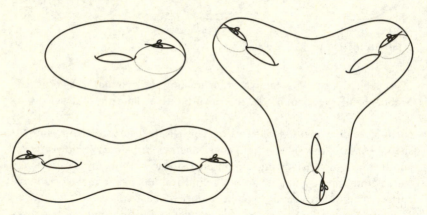

Figure 16.17. Surfaces of genus 1, 2, and 3.

(see figure 16.17), but after this first cut, it is impossible to find another such closed curve. So the genus of a torus is 1.

The genus of a sphere with handles is simply the number of handles. A double torus has genus 2 and in general, a *g*-holed torus has genus *g*. The genus of a surface gives a rigorous way of defining Lhuilier's count of the number of "tunnels." We could define the genus for nonorientable surfaces, and some people do. However, because the genus is so intimately related to the number of holes of a torus, it is not usually used in the nonorientable case.

Just as we can create an array of orientable surfaces by adding handles, we can use a similar procedure to create nonorientable surfaces. In order to

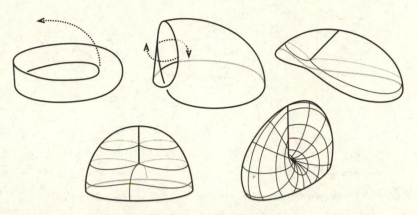

Figure 16.18. A Möbius band is the same as a cross cap.

understand this procedure, we must return to the Möbius band. One of its notable properties is its single boundary circle. The Möbius band is usually drawn so that this circle wraps around the twisted cylinder two times. Our aim is to manipulate the Möbius band so that its boundary circle appears like a regular circle, not one twisted twice. Clearly we need four dimensions to accomplish this feat of topological yoga.

In figure 16.18 we see the Möbius band deformed in this way. Notice that this figure passes through itself along an entire line segment. The self-intersection at the top of this Möbius band with the cusp at the top and the crisscrossing surface below it is often called a *Whitney umbrella*, named for the topologist Hassler Whitney. This strange presentation of the Möbius band is called a *cross cap*. The resemblance to a projective plane should be apparent, because a cross cap is simply a projective plane with a disk removed.

We can form nonorientable surfaces by attaching Möbius bands. To do so, remove a disk from the surface and glue the circular boundary of a Möbius band to the boundary of the hole. As we can see in figure 16.19, it is easier to visualize the gluing by replacing the typical Möbius band by a cross cap. We create a projective plane by adding one cross cap to a sphere. Stated another way, a projective plane is a Möbius band with a disk glued to its boundary.

It is not as easy to see, but a sphere with two cross caps is a Klein bottle. Equivalently, a Klein bottle is obtained by gluing two Möbius bands together along their boundaries. Thus, the limerick at the beginning of the chapter makes sense. Gluing more than two cross caps to a sphere yields even stranger nonorientable surfaces.

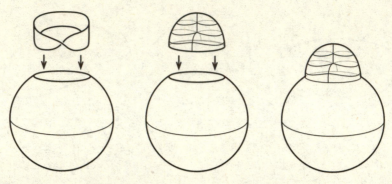

Figure 16.19. A sphere with a cross cap (creating a projective plane).

We now have two ways to construct orientable and nonorientable surfaces. In the next chapter we will investigate how to apply Euler's formula to these surfaces. We will also present the classification theorem for surfaces that states that every closed surface can be made by adding handles and cross caps to a sphere.

CHAPTER 17

ARE THEY THE SAME, OR ARE THEY DIFFERENT?

It was very often repeated that Geometry is the art of reasoning well on badly made figures; still these figures, not to mislead us, must satisfy certain conditions; the proportions can be grossly distorted, but the relative positions of the various parts should not be disrupted.
—Henri Poincaré in the introduction to *Analysis Situs*[1]

One of the most important recurring questions in mathematics is: are the two mathematical objects X and Y the same? In different contexts we have different criteria for what "the same" means. Often, when we say the same we mean equal, such as the expression $5 \cdot 4 + 6 - 2^3$ and the number 18, or the polynomials $x^2 + 3x + 2$ and $(x + 2)(x + 1)$. In other circumstances the same may not mean equal. For a sailor navigating by compass, two angles are the same if they differ by 360° (30° is the same as 390°). A geometer may say that two triangles are the same if they are congruent or perhaps if they are similar.

In topology we have a looser set of criteria for sameness than we do in geometry. This is where the rubber-sheet analogy comes into play. Intuitively, if one shape can be continuously deformed into the other, then they are the same. Bending, twisting, stretching, and squashing the shape does not change its topology. For example, the circle shown in figure 17.1 is the same as the tangle to its right. On the other hand, puncturing a shape, cutting it, or gluing it to itself will likely yield a shape that is topologically different. A circle is not the same as a circle glued to itself in the form of a figure eight.

In the first half of the nineteenth century mathematicians struggled to classify those polyhedra that satisfy Euler's formula—the so-called Eulerian polyhedra. We came to the vague understanding that all polyhedra

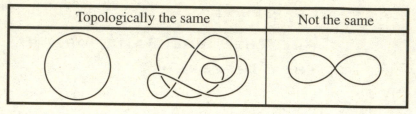

Topologically the same	Not the same

Figure 17.1. The tangle is topologically the same as a circle, but the figure eight is not.

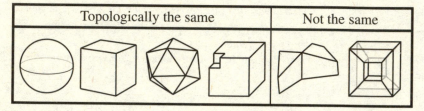

Topologically the same	Not the same

Figure 17.2. Polyhedra that are topologically the same as and different from a sphere.

that are "sphere-like" are Eulerian, whereas the bizarre exceptions of Lhuilier and Hessel are not. It turns out that Euler's formula applies to any polyhedron that is topologically the same as a sphere. A cube, any Platonic or Archimedean solid, and even certain nonconvex polyhedra can be deformed into a round, spherical ball (see figure 17.2). Non-Eulerian polyhedra, such as one formed by joining two polyhedra along an edge or one shaped like a torus, are not topologically the same as a sphere.

It may seem that this study of shapes is intuitive, but it is remarkable how often we stumble upon counterintuitive results. For instance, in figure 17.3 we begin with a double torus hanging on a clothesline with the string passing through one of the holes. By topological manipulations (and no cutting or gluing!) we achieve what may at first appear to be an impossibility—we transform it into a double torus hanging on a clothesline through both holes.

In chapter 16 we saw the difference between extrinsic and intrinsic dimension. We could use similar terminology in this context as well. The examples shown earlier in this chapter have what we might call the same extrinsic topology because one shape can be deformed into the other in 3-dimensional space. Mathematicians call two shapes with the same extrinsic topology *isotopic*. Isotopic is a valid choice for the definition of

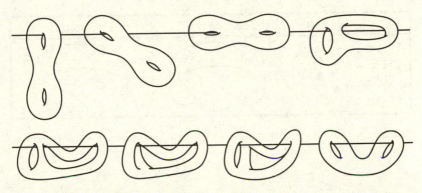

Figure 17.3. The double-torus clothesline trick.

Topologically the same		Not the same

Figure 17.4. Shapes that are topologically the same as and different from a torus.

topologically "the same," but it turns out that topologists want more freedom. We need a less restrictive definition for "the same."

For two shapes to be topologically the same, they must have the same intrinsic topology. If two surfaces are the same, then regardless how clever an ant living on the surface is, she will not be able to tell one from the other without leaving the surface. It is possible to find two surfaces that are the same, but for which one cannot be deformed into the other. Thus the rubber sheet analogy is not a perfect one.

To understand this new definition, we must revisit our cutting and gluing policy. While it is true that cutting and gluing usually change the topology of a surface, this is not always the case. An important exception is: cutting a shape, then gluing the severed pieces so that the cuts line up exactly as they did before. In this case the topology does not change. If we cut a torus around the tube to form a cylinder, tie the cylinder into a knot, and then reglue it (as in figure 17.4), the resulting shape is still topologically the same as a torus. Notice that this knotted torus cannot be obtained from the original torus simply through deformations in 3-dimensional

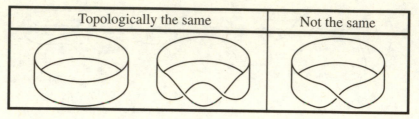

Topologically the same	Not the same

Figure 17.5. A band with two half-twists, but not a band with a single half-twist, is homeomorphic to a cylinder.

space—they are not isotopic. The intrinsic topology is the same, but the extrinsic topology is not. On the other hand, there is no way to cut, deform, and reglue a torus to obtain a double torus. A sufficiently clever ant will be able to prove that they are topologically different (and we will see how very soon).

Because it is outside the scope of this book, we omit the topologist's precise definition of "the same." In essence, two topological objects are the same if there is a one-to-one correspondence between the points in the two objects that preserves closeness—nearby points in one shape correspond to nearby points in the other. This notion of "the same" was introduced by Möbius, who called the correspondence an "elementary relationship."[2] Today such a correspondence is called a *homeomorphism*. Thus, in the lingo of the topologists, two shapes are the same whenever they are *homeomorphic*.

Consider the three loops of cloth from the Afghan bands magic trick in chapter 16. One has no twist, one has a single half-twist, and one has two half-twists. Clearly none have the same extrinsic topology. However, according to our rule of thumb, the third shape is homeomorphic to the untwisted cylindrical band, for if we cut the cylinder and gave two half-twists, the cut edges would line up correctly to be reglued (figure 17.5). We will call this third shape a twisted cylinder. This is not the case for the Möbius band. If we cut the cylinder and gave it a half-twist, the cut edges would not line up correctly. Similarly, despite the superficial resemblance between a Möbius band and a twisted cylinder, they are not homeomorphic.

Although our intuition tells us that the Möbius band is not homeomorphic to the cylinder (twisted or not), we have not given a proof. Although it seems unlikely, perhaps there is an elaborate cutting scheme that would turn one into the other. As we learned from the double-torus clothesline trick, we cannot always rely on our gut feeling, but in this case our intuition is correct—they are not homeomorphic.

A *topological invariant* is a property or mathematical entity associated with a surface that depends only on the topology of the surface. A topological invariant may take the form of a number, such as the number of boundary components. If two surfaces are homeomorphic, then they must have the same number of boundaries. In practice, the contrapositive is more useful: if two surfaces have a different number of boundaries, then they cannot be topologically the same. Because a cylinder has two boundary components and a Möbius band has one, they are not homeomorphic.

Intrinsic dimension is another topological invariant: it allows us to distinguish a (2-dimensional) sphere from a (1-dimensional) circle. We will discuss the notion of dimension more thoroughly in chapter 22.

Orientability is another topological invariant, or more specifically, a *topological property*. Two surfaces that are topologically the same are both orientable or both nonorientable. Said another way, if one surface is orientable and another is not, then they must not be homeomorphic. It is not difficult to see that a cylinder and a twisted cylinder are both orientable, but a Möbius band is not.

According to our cutting and gluing rules, a strip of paper glued together with an even number of half-twists is topologically the same as a cylinder and one glued with an odd number of half-twists is the same as a Möbius band. The strips with an even number of half-twists are orientable and have two boundary components, while those with an odd number of half-twists are nonorientable and have one boundary, and thus the strips are not homeomorphic to each other. Notice, by the way, that each twisted band has a mirror twin. When it comes time to twist and glue, there are two choices for twisting, a right-hand twist or a left-hand twist.

Orientablility, dimension, and the number of boundary components are three important topological invariants. Another topological invariant, arguably the most important one, is the quantity $V - E + F$. Given a surface S partitioned into V vertices, E edges, and F faces (of course, we still need to avoid ring-shaped faces), we define the *Euler number* of S to be $V - E + F$ (the Euler number is often called the *Euler characteristic*). It is customary to use the Greek letter *chi* to denote the Euler number, so $\chi(S) = V - E + F$.

> The Euler number is a topological invariant for surfaces.

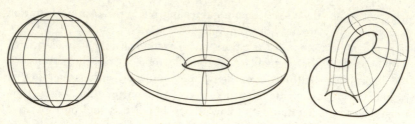

Figure 17.6. Partitions of the sphere, the torus, and the Klein bottle.

By saying that the Euler number is a topological invariant, we mean that each surface has its own Euler's formula. For example, the sphere in figure 17.6 has 62 vertices, 132 edges, and 72 faces, so its Euler number is

$$\chi(\text{sphere}) = 62 - 132 + 72 = 2.$$

As we know, this is true for any partition of a sphere or anything homeomorphic to a sphere.

The torus in figure 17.6 has 8 vertices, 16 edges, and 8 faces, so its Euler number is

$$\chi(\text{torus}) = 8 - 16 + 8 = 0.$$

Likewise, the Klein bottle in figure 17.6 has 8 vertices, 16 edges, and 8 faces, so

$$\chi(\text{Klein bottle}) = 8 - 16 + 8 = 0.$$

There are several steps involved in proving that the Euler number is a topological invariant. First we must show that any surface can be partitioned into finitely many vertices, edges, and faces. That is, there is no surface that is so bizarre that it has no finite partition (here is where we use the compactness assumption discussed in chapter 16—the Euclidean plane and the open unit disk have no finite partition, and are excluded). When we were working with polyhedra, we were given the partition—it was simply the vertices, edges, and faces of the polyhedron. A surface does not have a built-in partition. Surprisingly, it was not until 1924 that the first proof appeared that is it possible to partition every surface into vertices, edges, and faces.[3]

Next we must argue that the Euler number is independent of the choice of partition. It is not difficult to see that adding vertices and edges to a partition will not change $V - E + F$. So we ask, given two partitions of a surface, P and P', is it possible to add vertices and edges to P and P' so that the two partitions have the same number of vertices, edges, triangular faces, square faces, pentagonal faces, and so on, and that their placement relative to each other is the same? This problem was recognized early and was nicknamed the *Hauptvermutung*, which is a short for "the main conjecture of combinatorial topology." Its proof came late—not until 1943[4]—and as we will see in chapter 23, things are not so nice in higher dimensions. Because the *Hauptvermutung* is true for every surface, the Euler number is independent of the choice of partition.

Finally, we must show that two homeomorphic surfaces have the same Euler number. If S and S' are homeomorphic surfaces and P is a partition of S, then because the homeomorphism gives a one-to-one correspondence between S and S', we may use the homeomorphism to transfer the partition P over to a partition of S'. Clearly, then, $\chi(S) = \chi(S')$. Thus, we have completed the sketch of the proof of our theorem—that the Euler number is a topological invariant.

One of the biggest challenges in the study of Euler's formula for polyhedra was to understand the effect of "tunnels" on $V - E + F$. Lhuilier and Hessel both asserted that if a polyhedron had g tunnels, then $V - E + F = 2 - 2g$. Using our modern terminology, they asserted that the Euler number is $2 - 2g$. The problem was that they did not define tunnel. Instead of tunnels, we now use handles (in the sense of chapter 16) to describe these topological features. It is interesting that they focused on the holes in the figures, and we now focus on the handles that bound these holes.

Let us examine the effect on the Euler number of adding a handle to a sphere. We must remove two disks from the sphere so that we can attach the handles in their place. We may as well take these two disks to be triangular faces (see figure 17.7). If the partition does not have triangular faces, subdivide a face into triangles. We know that the Euler number of the sphere is 2, and the handle is a cylinder, which has Euler number 0 (the simplest partition has $V = 2$, $E = 3$, and $F = 1$). So, before any cutting and gluing, we have

$$V - E + F = \chi(\text{sphere}) + \chi(\text{handle}) = 2 + 0 = 2.$$

When we cut away the two triangles we lose two faces. When we glue the handle to the sphere, six pairs of edges are joined. Thus those twelve edges become six edges. Similarly, the twelve vertices become six vertices. After cutting and gluing, V and E both decrease by six and F decreases by two, so $V - E + F$ decreases by two. So,

$$V - E + F = \chi(\text{sphere}) + \chi(\text{handle}) - 2 = 2 - 2 = 0.$$

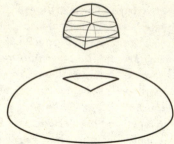

Figure 17.7. Adding a handle and a cross cap to a sphere.

Of course, we know that a sphere with a handle is a torus, so this is not surprising.

By the same reasoning, each time we add a handle to a surface, the Euler number decreases by 2. Thus we have a proof of Lhuilier's observation:

$$\chi(\text{sphere with } g \text{ handles}) = 2 - 2g.$$

Likewise, we can compute the effect of adding a cross cap. Notice that $\chi(\text{cross cap}) = \chi(\text{Möbius band}) = 0$ (like a cylinder we can have $V = 2$, $E = 3$, and $F = 1$). Because each cross cap has a single circle as a boundary, we must remove only one face from the sphere for each cross cap that we add. Again, assume that this face is a triangle. Using the same rationale as above, when we add a cross cap, the numbers of edges and vertices decrease by 3 and the number of faces decreases by 1. So after adding a cross cap, $V - E + F$ decreases by 1. For a sphere with one cross cap we have

$$V - E + F = \chi(\text{sphere}) + \chi(\text{cross cap}) - 1 = 1.$$

We conclude that the Euler number of the projective plane (a sphere with one cross cap) is 1. When we add c cross caps we obtain

$$\chi(\text{sphere with } c \text{ cross caps}) = 2 - c.$$

We now know how to compute the Euler number of any surface that can be obtained from a sphere by adding handles and cross caps. An important question remains—are there any surfaces that cannot be obtained in this way? In other words, can we describe all of the possible surfaces in terms of handles and cross caps? Using mathematical jargon, we ask: is it possible to *classify* all surfaces?

In mathematics, classification theorems are typically difficult or impossible. It is no wonder that Euler never completed his classification of polyhedra. However, occasionally it is possible to classify mathematical objects. After all, Theaetetus classified all regular polyhedra and Archimedes classified all semiregular polyhedra.

Remarkably, it is possible to classify surfaces (with and without boundary). Every closed surface is homeomorphic to a sphere with handles or a sphere with cross caps. That is, every orientable surface is topologically the same as a multi-holed torus, and every nonorientable surface is the same as a sphere with one or more Möbius bands attached to it. In fact, the theorem is stronger than this. If we are given some arbitrary closed surface and we know the Euler number and whether it is orientable or not, then we can identify the surface precisely.

CLASSIFICATION THEOREM FOR SURFACES

A closed surface is uniquely determined by its Euler number and its orientability. An orientable surface is homeomorphic to a sphere with g handles, for some $g \geq 0$. A nonorientable surface is homeomorphic to a sphere with c crosscaps, for some $c > 0$.

For example, suppose S is an orientable closed surface with Euler number -6. Since S is orientable we know that it is homeomorphic to a surface of genus g (a sphere with g handles), where $-6 = \chi(S) = 2 - 2g$. That is, S is homeomorphic to a 4-holed torus. Likewise, if T is a nonorientable closed surface with Euler number -4, then it is homeomorphic to a sphere with c cross caps, where $-4 = \chi(T) = 2 - c$. In other words, T is homeomorphic to a sphere with 6 cross caps.

A similar classification theorem holds for surfaces with boundary. Any surface with boundary is one of these standard surfaces with one or more disks removed. The Euler number, the orientability, and the number of boundary components determine the surface precisely. The only orientable surface with Euler number 0 and two boundary components is a cylinder, the only nonorientable surface with Euler number 0 and one boundary component is a Möbius band, and so on (see table 17.1).

In a sense, it was Bernhard Riemann (1826–1866) in the 1850s who began the classification process. Riemann was one of the preeminent mathematicians of the nineteenth century. He earned his doctoral degree at Göttingen under Gauss at the very end of Gauss's career. At that time Göttingen was not a center of mathematics in Germany (Gauss taught only introductory courses), so Riemann did much of his graduate work at the University of Berlin.

TABLE 17.1:
The Euler number, orientability, and number of boundaries for different surfaces.

Surface S	$\chi(S)$	Orientable	Boundaries
Sphere	2	Yes	0
Torus	0	Yes	0
2-holed torus	−2	Yes	0
g-holed torus	$2 - 2g$	Yes	0
Disk	1	Yes	1
Cylinder/annulus	0	Yes	2
Klein bottle	0	No	0
Projective plane	1	No	0
Sphere with c cross caps	$2 - c$	No	0
Möbius band	0	No	1

Figure 17.8. Bernhard Riemann.

His brilliance was recognized early. Gauss, who did not give praise lightly, was impressed by Riemann's first public lecture, given in 1854. Freudenthal describes this lecture as:

One of the highlights in the history of mathematics: young, timid Riemann lecturing to the aged, legendary Gauss, who would not live past the next spring, on consequences of ideas the old man must have recognized as his own and which he had long secretly cultivated. W. Weber recounts how perplexed Gauss was, and how with unusual emotion he praised Riemann's profundity on their way home.[5]

Much of Riemann's work was in complex analysis—the study of complex numbers and complex functions. A complex number has the form $a + bi$, where a and b are real numbers and $i = \sqrt{-1}$. It was in a quest for a complete understanding of complex functions that most of his work originated—function theory, geometry, partial differential equations, and topology. Some of his work appeared posthumously, including his treatment of integration, which is now a staple of every introductory calculus course. It is unfortunate that the life of this original thinker was cut short by tuberculosis at the age of forty.

Riemann's interest in surfaces did not come from the theory of poly-hedra, but from complex analysis. Because the complex numbers have two degrees of freedom (a and b), the collection of complex numbers forms a two dimensional plane—it looks like the Euclidean plane except one axis is real and the other one is imaginary.

Riemann was studying multivalued complex functions. For example, consider the function $f(z) = \sqrt[4]{z}$. What is the value of $f(16)$? It is a number w with the property that $w^4 = 16$. It is not difficult to see that there are four such complex numbers, 2, -2, $2i$, and $-2i$. Thus, a single input has more than one output. One way to think of this is that the graph of the function has several layers, or branches. Riemann cleverly interpreted this layered graph as a surface, now called a *Riemann surface*. A Riemann surface can have quite interesting topology, but is always orientable.

It was in this way that Riemann began his investigations of topology. He introduced the genus of a surface (and the related concept of *connectivity*, which we will discuss in chapter 22). He grouped together orientable surfaces according to their genus, and he recognized intuitively that two topologically equivalent surfaces must have the same genus.[6] Despite this grouping of like surfaces, he did not recognize the converse: that two surfaces with the same genus are topologically the same.

Möbius was the first person to state and prove the classification theorem for orientable surfaces. Möbius had a tool that Riemann did not have. In 1863 he developed the idea of an elementary relationship (essentially what we call a homeomorphism). So he was able to say with some precision what it meant for two surfaces to be the same. Möbius showed that every orientable surface was topologically the same as one of the normal forms shown in figure 17.9—a sphere, a torus, a double torus, and so on.[7]

In 1866 Camille Jordan proved that any two orientable surfaces with boundary are homeomorphic if and only if they have the same genus and the same number of boundary components.[8] The first complete statement and proof of the classification theorem, including nonorientable surfaces,

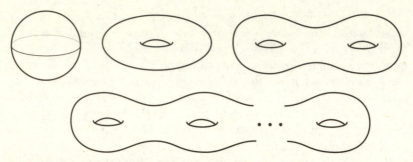

Figure 17.9. Möbius's normal forms for orientable surfaces.

was due to Dyck in 1888.[9] However this was still before the modern definitions of surface and homeomorphism. The first truly rigorous proof of the classification theorem is due to Max Dehn (1878–1952) and Poul Heegaard (1871–1948) in 1907.[10]

We will not prove the classification theorem, but there are a number of readable demonstrations. Some entail building up the surface to obtain a sphere with handles and cross caps. For instance, John Conway's ZIP proof ("zero irrelevancy proof") starts with a pile of triangles—the unassembled puzzle pieces of a triangulated surface. As each new triangle is zipped onto the ever-expanding surface, it remains a sphere with handles, cross caps, and boundary.[11] Other proofs work in reverse—they begin with a surface and repeatedly cut out cylinders and Möbius bands (i.e., handles and cross caps), filling in the holes with disks at each stage until arriving at a sphere.

At first glance it may seem that it would be easy to determine the genus of an orientable surface—after all, it is just a sphere with handles. However, they do not always look like one of Möbius's normal forms. For example, the first surface in figure 17.10 is an example of a sphere with 4 handles—it is homeomorphic to a 4-holed torus.

The classification theorem says that every surface is homeomorphic to a sphere with handles or a sphere with cross caps. It does not say anything about mixing the two. For instance, the second picture in figure 17.10 is a sphere with one handle and one cross cap. How does this fit into the classification? According to our calculations above, the sphere has Euler number 2, adding a handle decreases it by 2, and adding a cross cap decreases it by 1. So the Euler number of this surface is −1. Because of the presence of the cross cap, we know that the surface is nonorientable. According to the classification theorem, it is homeomorphic to a sphere with three cross caps, a surface known as *Dyck's surface*.[12]

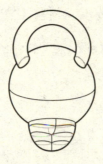

Figure 17.10. Unusual surfaces.

A quick inspection of the third surface in figure 17.10 shows that it is two-sided (orientable) and has only one boundary component. Interestingly, the boundary itself forms what is called a trefoil knot. As we will see in the next chapter, any knot can be obtained as the boundary of an orientable surface with one boundary component. By forming a partition of this surface and counting vertices, edges, and faces, we find that the Euler number is −1. By the classification theorem for surfaces with boundary, this surface is homeomorphic to a torus with a disk removed.

As a final example we return to the great icosahedron and the great dodecahedron—the Kepler-Poinsot polyhedra with trangular and pentagonal faces (see chapter 15). Although they do not look like it, they are orientable surfaces (with self-intersections in 3-dimensional space). The Euler number of the great icosahedron is 2, so it is homeomorphic to a sphere, and the Euler number of the great dodecahedron is −6, so it is homeomorphic to a 4-holed torus.

CHAPTER 18

A KNOTTY PROBLEM

O Time, thou must untangle this, not I.
It is too hard a knot for me t'untie.
—William Shakespeare, *Twelfth Night*[1]

One of the earliest topological investigations was the study of *knots*. We are all familiar with knots. They keep our boats secured to shore, our shoes snug on our feet, and the cables and wires hopelessly tangled behind our computers. These are not, strictly speaking, mathematical knots. A mathematical knot has no free ends; it is a topological circle living in 3-dimensional Euclidean space. (To turn an electrical extension cord into a mathematical knot, simply plug the two ends together.)

In figure 18.1 we see the projections of six mathematical knots: the *unknot, trefoil knot, figure eight knot, pentafoil knot, gingerbread man knot* (for lack of an accepted name), and *square knot*.

In the previous chapter we stressed that topologists are usually interested in the intrinsic properties of topological objects, not the extrinsic properties. Knot theory is one notable exception. What is interesting about a knot is how the circle is placed in space—its extrinsic configuration. Intrinsically, they are all identical—every knot is homeomorphic to a circle. So in the study of knots, "the same" does not mean homeomorphic. Instead, two knots are the same if one can be continuously deformed into the other. That is, two knots are the same if there is an isotopy between them. The first three knots in figure 18.2 are isotopic (they are all the same as the unknot) and the last two are the isotopic (they are both the same as the trefoil knot), but as we will see, the unknot is not isotopic to the trefoil knot.

A main goal of knot theory is to classify knots. Just as we did for surfaces, we would like to find tools that allow us to tell whether two knots are the same or different. Ideally, like surfaces, we dream about producing an exhaustive and duplicate-free list of all knots. Right now no such complete

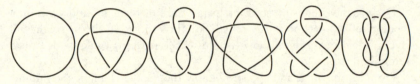

Figure 18.1. The unknot, trefoil knot, figure eight knot, pentafoil knot, gingerbread man knot, and square knot.

Figure 18.2. Three projections of the unknot and two projections of the trefoil knot.

list exists, but much has been accomplished in this direction. The meager goal of this chapter is to develop enough tools to prove that the knots in figure 18.1 are all different. One of these tools requires the classification of surfaces and the Euler number.

The study and use of knots is as old as humankind. A vast array of bends, hitches, loops, splices, and nooses have been discovered for every conceivable use. In many cultures knots and their projections were common themes for jewelry and artwork. They were also important in the production of fabrics, for what is a piece of cloth if not a giant knot? In comparison, the mathematical study of knots is considerably younger; the first mathematical investigation dates back to the eighteenth century.

The topological significance of knots was recognized by Alexandre-Théophile Vandermonde (1735–1796) in 1771, only thirty-five years after Euler's bridges of Königsberg paper. Vandermonde's short paper "Remarques sur les problèmes de situation" ("Remarks on the problems of position") began as follows:

> Whatever the twists and turns of a system of threads in space, one can always obtain an expression for the calculation of its dimensions, but the expression will be of little use in practice. The craftsman who fashions a braid, a net, or some knots will be concerned, not with questions of measurement, but with those of position; what he sees there is the manner in which the threads are interlaced.[2]

Despite the promising start, his paper did not focus on knots but on a topological approach to the so-called "knight's tours" in chess. Yet, he did give a brief description of how certain textile patterns could be described symbolically.

From his drawings and notes, we know that Gauss thought about knots as early as 1794, but alas, he never published anything on the subject. In one handwritten gem from 1833 he gave a double integral that could be used to compute the *linking number* of two closed curves—a topological quantity that measures how many times the curves wind around each other.[3]

Perhaps it is not surprising, then, that Listing, one of Gauss's students, truly began the mathematical study of knots. His contributions can be found in his 1847 monograph *Topologie*,[4] the oft-cited treasure trove of topological curiosities. Although Listing did not propose the classification of all knots, he was clearly interested in finding techniques for differentiating two knots. For instance, he stated that the trefoil knot and its mirror image are not the same knot. Just as Listing's treatment of the Möbius band was ignored, so was his study of knots. Ultimately, the rise of knot theory was due not to Listing, but to two Scottish physicists working on a new atomic theory.

In 1867 William Thomson (1824–1907) asserted that atoms were formed by vortices, or knots, in the ether. Perhaps better known as Lord Kelvin, Thomson is also responsible for the absolute temperature scale and for helping design the first transatlantic telegraph cable (he was knighted for this work). According to Kelvin each atom corresponded to a different knot or a linked collection of knots, and that the stability of an atom was due to the stability of the knot under topological deformation. This ingenious but misguided belief prevailed for two decades.

Thomson's atomic theory led his friend Peter Guthrie Tait to commence the classification of knots. In 1877 Tait began a tabulation of all knots—in his eyes he was creating a periodic table of elements. Eventually this view of chemistry was abandoned, but Tait continued his study. By 1900 he and the Indian-born American mathematician Charles Newton Little (1858–1923) had given an almost complete account of knots with ten or fewer crossings (by crossings, we mean in a given planar drawing—we will say more about this terminology shortly).

Tait primarily used his remarkable intuition to classify knots. In the years afterward, mathematicians devised a myriad of ingenious tools for rigorously differentiating knots. Most of these tools are *knot invariants*. In

Figure 18.3. William Thomson (Lord Kelvin) and Peter Guthrie Tait.

chapter 17 we discussed topological invariants for surfaces. Knot invariants play the same role for knots. A knot invariant is a number, or other entity, that is associated to a knot. If the invariants of two knots are different, then they must be different knots.

There are many knot invariants, and some are quite easy to describe. In this chapter we will introduce a few knot invariants, including one that is related to surfaces and the Euler number.

A knot is a circle, and circles can be found on the boundaries of surfaces. Surprisingly, it is possible to find surfaces with knotted boundaries. In figure 18.4 we see that the unknot is the boundary of a disk (no surprise here) and the trefoil knot is the boundary of a Möbius band with three half-twists. We saw another example of a surface with a trefoil knot boundary in figure 17.10.

Remarkably, not only is it possible to find a surface with a knotted boundary, it turns out that *every* knot can be realized as the boundary of a surface.

As a fun experiment, try making surfaces with knotted boundaries out of soap bubbles. To do so, make a knot out of stiff wire (a coat hanger will suffice for small knots, although it is too stiff and not long enough to make

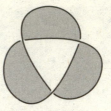

Figure 18.4. The unknot is the boundary of a disk and the trefoil knot is the boundary of a Möbius band with three half-twists.

Figure 18.5. Herbert Seifert.

elaborate knots), then dip it into a bubble solution. Poke holes as needed to form a single surface.*

The trefoil in figure 18.4 is the boundary of a nonorientable surface (remember that in 3-dimensional space, nonorientable and one-sided are synonymous). It is possible to avoid this situation—given any knot, we are able to construct an orientable surface with this knot as its boundary. Such a surface is called a *Seifert surface*, named for Herbert Seifert (1907–1996). Perhaps as surprising as this theorem is the simplicity of constructing such

---

* To make durable bubbles that can span large distances, we recommend mixing 1 gallon of water, 2/3 cup of dish soap, and 1 tablespoon of glycerin (which can be found at any drugstore). For best results, let the solution sit for a while before use.

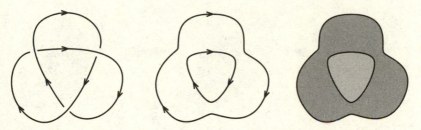

Figure 18.6. Seifert circles for a trefoil knot and the corresponding disks.

surfaces. We will now give Seifert's elegant algorithm, which he discovered in 1934.[5]

We illustrate the algorithm for a trefoil knot. To begin, pick one of the two possible orientations for the knot. That is, pick a direction of travel along the knot. Next, project the knot onto the plane. Almost any projection is allowable. We want to avoid "bad" projections such as those having three strands crossing at the same point or having two sections of string project on top of each other for more than a point, but otherwise the projection can be as complicated as we want.

Next, use this projection to create a collection of circles called *Seifert circles*. Start tracing the knot following the orientation of the knot. At each crossing, change to the other strand, but still follow the direction of orientation. A circle is obtained when the tracing returns to the starting point (see figure 18.6). Repeat on all untraced strands of the knot. Given these Seifert circles, create disks having these circles as boundaries. Depending on the projection, some of the circles may be nested inside each other. In this case the disks will sit atop one another (as is the case with the trefoil in figure 18.6).

Now, join the disks together by attaching rectangular bands, each with a half-twist. Specifically, at each location where there had been a crossing, attach a twisted band with the direction of the twist determined by the original crossing (see figure 18.7). Although it is not immediately obvious, it is not too difficult to prove that this procedure always produces an orientable surface with boundary.

We complete the construction of the Seifert surface for the trefoil knot in figure 18.8. In figure 18.9 we repeat the process for the square knot. This surface is obtained from three disks and six bands.

According to the powerful classification theorem for surfaces, we "know" every possible surface. A Seifert surface is an orientable surface with one boundary component. Therefore, it must be homeomorphic to a

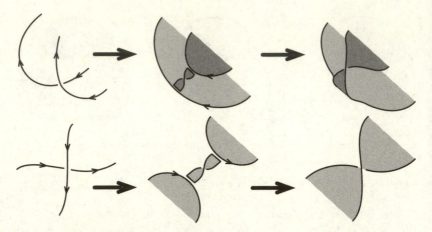

Figure 18.7. Attaching a band with a twist.

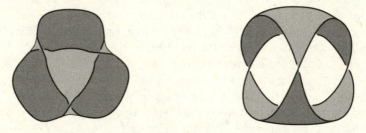

Figure 18.8. A Seifert surface for a trefoil knot.

Figure 18.9. A Seifert surface for a square knot.

sphere or g-holed torus with a disk removed. Now we truly see the power of the classification theorem, for Seifert surfaces do not look like punctured tori. Theoretically, we could glue a disk onto the boundary of one of these Seifert surfaces and obtain a closed surface, but this gluing would require four dimensions.

Because we know that a Seifert surface is orientable and has one boundary, we need only know its Euler number to classify it. Suppose S is a Seifert surface constructed from d disks and b bands. Since the Euler number of a disk is 1 (and thus the Euler number of d disjoint disks is d), it suffices to determine the effect of adding a band to a surface.

Suppose we attach both ends of a band to a surface with boundary (that may not be connected). In doing so we add one face, two edges and no vertices to the surface. According to the familiar alternating sum of the Euler number, adding the band decreases the Euler number by 1. So, adding b bands decreases it by b.

> A Seifert surface S constructed from d disks and b bands has Euler number $\chi(S) = d - b$.

We constructed the Seifert surface for the trefoil knot (figure 18.8) using two disks and three bands. So its Euler number is -1. Similarly, the Seifert surface for the square knot is made from three disks and six bands, so its Euler number is -3.

If we were to glue a disk to the boundary of a Seifert surface, we would obtain a closed surface of genus g. Doing so adds a face, so this closed surface would have an Euler number one larger than that of the Seifert surface. The Seifert surface for the trefoil knot shown in figure 18.8 has Euler number -1. If we attach a disk to the boundary, then the resulting surface has Euler number 0. It must be a torus—a surface of genus 1. We say that the Seifert surface has genus 1.

We can apply this same logic to any Seifert surface made from d disks and b bands. By attaching a disk to the boundary, we obtain a closed, orientable surface S with Euler number $\chi(S) = d - b + 1$. S is a surface of genus g, and $\chi(S) = 2 - 2g = d - b + 1$. Solving for g we find the genus.

> A Seifert surface made from d disks and b bands has genus $g = (1 - d + b)/2$.

The Seifert surface for the square knot has genus $g = (1 - 3 + 6)/2 = 2$. It is a 2-holed torus with a disk removed. In figure 18.10 we see Seifert surfaces for the pentafoil, figure eight, and gingerbread man knots. The Seifert surface for the pentafoil knot is constructed from two disks and five bands. So, its genus is $(1 - 2 + 5)/2 = 2$. The Seifert surface for the

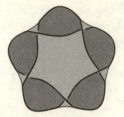  

Figure 18.10. Seifert surfaces for the pentafoil, figure eight, and gingerbread man knots.

figure eight knot is made from three disks and four bands, and has genus $(1 - 3 + 4)/2 = 1$. The Seifert surface for the gingerbread man knot, made from three disks and six bands, has genus $(1 - 3 + 6)/2 = 2$.

It would be nice if the genus of a Seifert surface was a knot invariant. The problem is that a given knot may have more than one topologically distinct Seifert surface (all it would take is to choose a different projection of the knot at the beginning of the process). However, we do not have to abandon this idea entirely. We may define the *genus* of knot to be the smallest genus of all possible Seifert surfaces. We denote the genus of a knot K by $g(K)$.

The unknot bounds a disk, which is a sphere with a disk removed, so it has genus 0. It is the only knot that bounds a disk, so the genus is a positive number for every knot that is not the unknot.

This definition is somewhat frustrating. While the genus is a perfectly valid knot invariant, in practice it is not easy to compute. We constructed a Seifert surface for the gingerbread man knot and found its genus to be 2. Is that the genus of the knot? Maybe, maybe not. Perhaps it is possible to find a different Seifert surface for this knot with genus 1. It is not clear how to prove that no such Seifert surface exists.

The good news is that we can easily compute the genus for a broad class of knots called alternating knots. Trace the projection of the figure eight knot in figure 18.1 with your finger and observe the behavior at the crossings. The string goes over, under, over, under, over, under, over, and under—it alternates throughout the projection. This projection is called alternating. The projections of the trefoil, pentafoil, and gingerbread man knots are also alternating, but the projection of the square knot is not. A knot is called *alternating* if it has some alternating projection. The given projection of the square knot is not alternating, but this does not mean that it is not an alternating knot, for it may have another projection that *is* alternating.

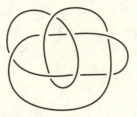

Figure 18.11. An 8-crossing knot that is not alternating.

All of the simplest knots are alternating. Every knot with seven or fewer crossings is alternating, and only three of the 8-crossing knots are not alternating (one is shown in figure 18.11). However, as the number of crossings increases, the relative number of alternating knots drops. Of the 2,404 prime knots (we will define prime shortly) with 12 or fewer crossings, 63% are alternating. Of the roughly 1.7 million prime knots with 16 or fewer crossings, only 29% are alternating.[6]

What saves us is a theorem proved at the end of the 1950s by Richard H. Crowell and Kunio Murasugi.[7]

> The Seifert surface obtained from an alternating projection is guaranteed to have the minimum genus.

In other words, because the trefoil, figure eight, pentafoil, and gingerbread man knots are alternating, we can say with certainty that their genuses are 1, 1, 2, and 2, respectively. So, none of the four knots is the unknot, and the trefoil and figure eight knots are different from the pentafoil and gingerbread man knots. At this point the reader should be able to prove that the two knots in the introduction (figure I.5) are different.

We now take a brief diversion to determine the genus of the square knot. Prime numbers are the basic building blocks of the positive integers. A number $p > 1$ is prime if whenever p is the product of two positive integers m and n, either $m = 1$ or $n = 1$; otherwise it is composite. In a similar manner we define prime knots, the basic building blocks for all knots. In order to do so we need a way to "multiply" knots.

Given knots K and L, the *product* of K and L, denoted $K \# L$, is formed as follows. Place the projections of K and L next to (but not overlapping) one another. Cut an outside strand of both knots and join the four ends together without introducing any new crossings. In figure 18.12 we see

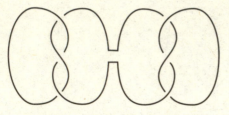 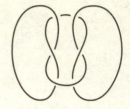

Figure 18.12. The product of a trefoil knot and its mirror image is a square knot.

that the square knot is the product of a trefoil knot and its mirror image (the product of two identical trefoil knots is a granny knot).

A knot M is *prime* if whenever $M = K\#L$, either K or L is the unknot.[†] In other words, a knot is prime if it cannot be written as the product of two nontrivial knots. A nontrivial knot that is not prime is called *composite*. Clearly, primality is a knot invariant. We have shown that the square knot is a composite knot. We will not prove it, but the trefoil, figure eight, pentafoil, and gingerbread man knots are all prime, so none of them are isotopic to the square knot.

Suppose we know the genuses of the knots K and L. Is it easy to determine the genus of $K\#L$? Suppose S_K and S_L are Seifert surfaces for K and L of minimal genus. Using the same projections of K and L, form $K\#L$ and its corresponding Seifert surface $S_{K\#L}$. It is not difficult to see that if S_K is formed from d_K disks and b_K bands and S_L is formed from d_L disks and b_L bands, then $S_{K\#L}$ is formed from $d_K + d_L - 1$ disks and $b_K + b_L$ bands. So, the genus of $S_{K\#L}$ is

$$\tfrac{1}{2}[1 - (d_K + d_L - 1) + (b_K + b_L)] = \tfrac{1}{2}(1 - d_K + b_K) + \tfrac{1}{2}(1 - d_L + b_L)$$

$$= g(K) + g(L).$$

The problem is, we do not know if $S_{K\#L}$ has the minimal genus for the knot $K\#L$. So all we can conclude is that $g(K\#L) \leq g(K) + g(L)$. We omit the proof, but in fact $S_{K\#L}$ does have the minimal genus. Thus, the genus is additive.

> For any knots K and L, $g(K\#L) = g(K) + g(L)$.

[†] The unknot satisfies this definition of primality, but just as 1 is not considered a prime number, the unknot is not considered a prime knot.

This formula enables us to compute the genus of the square knot:

$$g(\text{square knot}) = g(\text{trefoil knot}) + g(\text{trefoil knot}) = 1 + 1 = 2.$$

An interesting consequence of this formula is that if either K or L is not the unknot, then $K \# L$ is not the unknot. This follows because if either $g(K) \neq 0$ or $g(L) \neq 0$, then $g(K \# L) \neq 0$. Here is a concrete way of understanding this: if your shoelaces are knotted, then it is impossible to take the two free ends of the laces and tie a knot that unties the knot in your laces. Knots do not have "inverse knots" that untie them.

It is worth remarking that if K and L are alternating knots, then $K \# L$ is an alternating knot (can you see why this is true?). So the square knot, which we have shown without an alternating projection, has an alternating projection.

While the genus of a knot enables us to differentiate many knots, it is not a complete invariant—just because two knots have the same genus does not mean that they are the same knot. For example, the genuses of the trefoil knot and the figure eight knot are the same. So either these two knots are the same (which they are not), or we need to find a different method to distinguish them. Likewise, the genuses of the pentafoil, the gingerbread man, and the square knot are the same.

In the remainder of this chapter we introduce two more knot invariants that enable us to differentiate the remaining knots. They are just a small sampling of the many known knot invariants.

The first of these invariants is called colorability. To test for colorability, we draw the projection of a knot using three crayons of different colors. The knot is *colorable* if at each crossing only one color appears or if all three colors appear. Moreover, we insist that the entire projection not be just one color. It is not too difficult to prove that colorability is a knot invariant, but we will skip the proof. In particular, colorability does not depend on the choice of the projection.

In figure 18.13 we see that the trefoil is colorable (we used black, grey, and "dashed" as our three colors). However, a little experimentation shows that the figure eight is not colorable. In the example shown in figure 18.13 we obeyed the rules and colored the first three strands. At this point we run into trouble with the top strand. Each crossing would force this strand to be a different color. There is no color that we can use to color this knot correctly. So the trefoil and the figure eight knots are distinct.

We leave it to the reader to show that the square knot is colorable, but the pentafoil and the gingerbread man knots are not. Thus, we have further

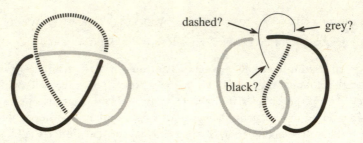

Figure 18.13. The trefoil is colorable and the figure eight is not.

TABLE 18.1:
The number of prime knots with a given crossing number.

$c(K)$	3	4	5	6	7	8	9	10	11	12	13	14
Knots	1	1	2	3	7	21	49	165	552	2176	9988	46972

proof that the square knot is distinct from the pentafoil and gingerbread man knots.

Using primality, genus, and colorability, we are able to differentiate all of our knots except the pentafoil and the gingerbread man knots. Both knots are prime, have genus 2, and are not colorable. To prove that they are distinct we must introduce one more knot invariant: the *crossing number*.

The crossing number of a knot is the fewest number of crossings of all projections of the knot. We will write $c(K)$ for the crossing number of a knot K. The usual projection of the unknot has no crossings, so it has crossing number 0. We know that the trefoil knot and the unknot are different, and we have a projection of the trefoil with 3 crossings. Any knot with zero, one, or two crossings is the unknot, so the crossing number of the trefoil is 3.

Knots are often grouped together by crossing number. There is not a lot of diversity among knots of few crossings. As we see in table 18.1, the trefoil is the only knot with crossing number 3 (not double-counting a knot and its mirror image), and there are only seven prime knots with six or fewer crossings. As the crossing number increases, the variety of knots increases rapidly.[8]

Like the genus, and for exactly the same reason, the crossing number is often difficult to work with. Given a projection, it is easy to count the number of crossings. However, we cannot guarantee that there is no other projection with fewer crossings. If we have a projection of a knot K with

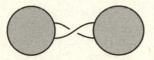

Figure 18.14. A nonessential crossing.

n crossings, all we can say is that $c(K) \leq n$. Fortunately, just like the genus, the crossing number is easy to compute for alternating knots.

A century ago Tait conjectured that a reduced alternating projection of a knot exhibits the minimum number of crossings. Here "reduced" means that before counting the number of crossings, we remove all nonessential crossings, such as the crossing in figure 18.14. Such a crossing can be removed by simply giving part of the knot a half-twist. Once these are removed, Tait conjectured, the number of crossings is the minimum. Tait's conjecture remained open for many years, but was proved independently and almost simultaneously by Louis Kauffman, Kunio Murasugi, and Morwen Thistlethwaite in the mid-1980s.[9]

> A reduced alternating projection of a knot exhibits the minimum number of crossings.

This theorem enables us to compute the crossing number of any alternating knot with ease. Because our projections of the trefoil, figure eight, pentafoil, and gingerbread man knots are already reduced and alternating, it is trivial to determine their crossing numbers. They are 3, 4, 5, and 6, respectively. So with this single invariant we conclude that all of these knots are distinct—including the pentafoil and the gingerbread man knots.

It is worth asking how crossing numbers are related to products of knots. Is there a nice relationship between $c(K)$, $c(L)$, and $c(K\#L)$? If K and L are both alternating, then $K\#L$ is alternating. Moreover, by being careful we can take reduced alternating projections of K and L and join them so that the resulting projection of $K\#L$ is also reduced and alternating (that is not what we did for the square knot). Consequently, in this special case the crossing number is additive.

> If K and L are alternating knots, then $c(K\#L) = c(K) + c(L)$.

For instance, $c(\text{square knot}) = c(\text{trefoil}) + c(\text{trefoil}) = 3 + 3 = 6$.

TABLE 18.2:
A summary of the properties of our knots.

Knot	Prime	Crossing no.	Genus	Colorable
Unknot	no	0	0	yes
Trefoil knot	yes	3	1	yes
Figure eight knot	yes	4	1	no
Pentafoil knot	yes	5	2	no
Gingerbread man knot	yes	6	2	no
Square knot	no	6	2	yes

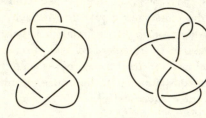

Figure 18.15. Are the gingerbread man and 6_3 knots the same?

Is the crossing number additive for all knots, just like the genus is? It is an old conjecture that this equality holds for all knots. Surprisingly, no one has discovered a proof, nor an example showing that it is false!

In this chapter we introduced a few of the many important knot invariants, and using them we were able to distinguish the six knots at the beginning of the chapter. We summarize our findings in table 18.2.

With these tools we can make a dent in the classification process. However, they only get us so far. The two projections in figure 18.15—the gingerbread man knot and the so-called 6_3 knot—are different knots, but our invariants cannot tell them apart. They are both prime, 6-crossing, alternating, not colorable, genus 2 knots. We need more tools to distinguish them.

Also, we have not presented any techniques for showing that two projections are the same knot. We devoted this chapter to showing that two projections are different knots. We encourage the reader to peruse the literature to investigate this interesting area of knot theory.

Mathematicians and scientists have a dysfunctional give-and-take relationship. Naively one would think that they work hand-in-hand. Scientists give problems to the mathematicians to solve and the mathematicians create theories that they hope will be useful to scientists.

The needs of scientists often do spur the creation of new mathematics, just as Kelvin's vortex model of atomic theory inspired knot theory. But mathematicians do not relish being servants of science. Even when a mathematical theory is born from a practical application, it quickly takes on a life of its own and is advanced for its own inherent properties. Theoretical mathematicians are a stubborn bunch who as a whole are more interested in beauty, truth, elegance, and grandeur than in applicability.

When Kelvin's atomic model was proved wrong, scientists lost interest in knots, but mathematicians continued studying them. Knot theory took on a life of its own as an area of pure mathematics. For most of the twentieth century it was of interest only to mathematicians. Knot theory remained an active research area with applications to other areas of pure mathematics, but not to science.

But even the most abstract and theoretical areas of mathematics can prove useful. Applicable mathematics often comes from decidedly non-applicable areas. The usefulness of a particular theory is often not clear for many years. No one could have predicted that the study of prime numbers would later enable us to encrypt credit-card information so that it can be sent safely across the Internet. Nineteenth-century mathematicians did not know that their work in non-Euclidean geometry would provide the foundation for Einstein's theory of general relativity.

Toward the end of the twentieth century the usefulness of knot theory reemerged in the natural sciences. Physicists, biologists, and chemists discovered that the mathematical theory of knots gave them insight into their fields. Whether it is the study of DNA or other large molecules, magnetic field lines, quantum field theory, or statistical mechanics, knot theory now plays an important role.

Mathematicians work in a store that makes and sells tools. Occasionally they take special orders from their scientific customers, but most of the day they toil away making elegant tools, the uses for which have not yet been invented. Scientists visit this tool shop and browse the shelves in hopes that one of the tools fits their needs. The knot-theory aisle of the store, long ignored by scientists, is now bustling with activity. In the next chapter we will see how the emerging ideas of topology and the Euler number created yet another unexpectedly useful tool for scientists.

CHAPTER 19

COMBING THE HAIR ON
A COCONUT

Let chaos storm!
Let cloud shapes swarm!
I wait for form.
—Robert Frost, "Pertinax"[1]

Many scientists use mathematics as a tool to predict behavior. A scientist may have an equation or a system of equations that describes the interactions of the quantities in their model. Then they use mathematics to draw conclusions from these equations.

Often the mathematical models are *differential equations*. These describe the rates of change of various quantities as a function of time. For example, an ecologist might use a system of differential equations to model the population dynamics of rabbits and foxes living in a wildlife refuge driven by their predator-prey relationship. When the number of rabbits is large, the foxes have an abundant food supply. As the foxes feed on the rabbits, the number of foxes increases and the number of rabbits decreases. Eventually, the foxes' food supply dwindles, causing their population to decrease. This, in turn, allows the rabbits to prosper again. We see this cyclic behavior illustrated visually in figure 19.1.

Differential equations are expressed as an algebraic relationship among variables and their derivatives. By a *solution* to a differential equation, we mean that for a given initial condition we can predict the future behavior of the system. In other words, if we know how many rabbits and foxes there are today, we can predict how many there will be a year from now. The curve in figure 19.1 is the plot of a solution curve. The arrows indicate the direction of positive time. The curve is plotted in the *phase space*, a topological object representing all possible values of the variables. In this case the phase space is the first quadrant of the plane (since the numbers of

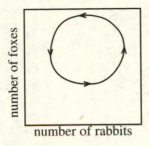

Figure 19.1. The predator-prey model.

rabbits and foxes must be nonnegative). In more exotic examples the phase space may be a more complicated topological shape.

Sometimes finding a particular solution to a differential equation is not sufficient for the scientist. It is often more important to draw qualitative conclusions. Does the system have an equilibrium—populations for which the death rate equals the birth rate for both species? Are there initial conditions that lead to extinction of either or both species? To a population explosion? Will the populations exhibit a cyclic behavior, or will they vary chaotically? Even though we may be able to solve a differential equation using calculus, it may not be easy to answer these important "global" questions.

In order to better understand the solutions of a system of differential equations, we may wish to come up with a more visual, geometric way to represent them. The two common techniques are to produce a *flow* or a *vector field* in the phase space. A flow, also known as a *continuous dynamical system*, associates to each point in the phase space the trajectory for the point to move. This trajectory is simply the solution curve for the differential equation. Several flow lines for the predator-prey model are shown in figure 19.2. They show that for any nonequilibrium pair of initial populations, the populations rise and fall cyclically.

Instead of expressing a differential equation algebraically or as a flow, we may describe it in terms of a vector field. In contrast to *scalar* quantities such as temperature, time, brightness, or mass, which can be described by the magnitude alone, *vector* quantities have a magnitude and a direction. In physics one might use a vector to describe velocity; the vector points in the direction of motion and the magnitude of the vector represents the speed at which the object is traveling. We could give other examples, but velocity is probably the most intuitive. In fact, if we think of a flow as the motion of

 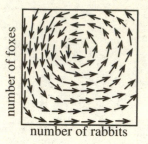

Figure 19.2. The associated flow and vector field for the predator-prey model.

particles, then the vectors in the vector field are the velocity vectors for the particles in the phase space.

In figure 19.2 we see that there is a single point of equilibrium—a point where the numbers of foxes and rabbits remain unchanged in time. The vector at this point of equilibrium has magnitude zero, so we say that vector field has a zero at this point. Equivalently, the flow here has a *fixed point* or *rest point*. Zeros of vector fields are often of great importance because they represent equilibrium points of the system.

We devote the rest of the chapter to an investigation of vector fields on surfaces and not to the differential equations from which they came. Our primary goal is to understand the relationship between the zeros of a vector field and the topology of the surface.

One of the easiest ways to obtain a vector field on a surface is to place the surface in 3-dimensional space and have the vectors "point downhill," and the steeper the descent, the longer the vector. This is called a *gradient vector field*. In figure 19.3 we show the gradient vector field for a sphere and a torus. One way to think about the flow associated to the vector field is to imagine the surface covered by molasses. The flow lines are simply the paths of the viscous syrup as it oozes down the surface.

As we see in figure 19.3, the gradient vector fields on the sphere and the torus have zeros—the sphere has two (one at the north pole and one at the south pole) and the torus has four (at the top and bottom of the torus and at the top and bottom of the hole). Must every vector field on the sphere have a zero? Must every vector field on the torus have a zero? If we can answer such a question "yes" for some surface S, then we have a powerful result. It says that any time the phase space of a system is S, it must have a point of equilibrium.

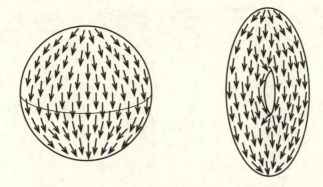

Figure 19.3. Gradient vector fields for a sphere and a torus.

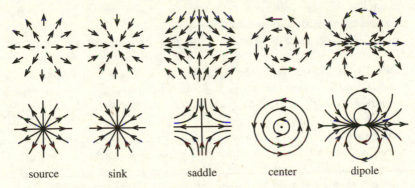

source   sink   saddle   center   dipole

Figure 19.4. Vector fields near a zero and their corresponding flows.

As a partial answer to these questions, we have the beautiful Poincaré-Hopf theorem. It establishes a surprising relationship between the zeros of a vector field and the Euler number. In order to understand the theorem we must look a little closer at zeros of vector fields.

Not all zeros are the same. In figure 19.4 we see five different zeros and their corresponding fixed points for the flow. The behaviors in the neighborhood of the zeros are all very different. A *source* repels all nearby points, a *sink* attracts them, a *saddle* does both, all points near a *center* flow around the fixed point, and points near a *dipole* flow away and then back again (resembling the magnetic field lines of a bar magnet).

Sources, sinks, and saddles often arise as zeros of the gradient flow. In figure 19.5 we see that a source appears at the top of an overturned bowl, a sink at the bottom of a bowl, and a saddle from the middle of a

Figure 19.5. A source, a sink, and a saddle in a gradient vector field.

saddle-shaped surface. Notice that in figure 19.3 the gradient flow on the sphere has one source and one sink, and the gradient flow of the torus has one source, one sink, and two saddles.

Our intuition tells us that sinks, saddles, and sources are somehow distinct and that there must be a rule that governs the number of each type of zero. We see from the gradient-flow example that a sphere can have a vector field with two zeros, but it is difficult to imagine a vector field on the sphere with one sink and one saddle. The key idea that we need to distinguish zeros is the *index*.

We compute the index of a zero as follows. Draw a small circle around the zero. The only conditions that this circle must satisfy are that (1) it must contain only one zero, and (2) it must be the boundary of a disk (for example, such a circle on torus cannot go around the tube or the central hole). Now place an imaginary dial at some point on the circle. The hand of the dial should point in the same direction as the vector field. (If the vector field represented the direction of a magnetic field, then we could use a compass as the dial.) As we move the dial around the circle, the hand of the dial turns. Move the dial around the circle one time counterclockwise. Each time the hand turns once counterclockwise, add one to the index, and each time it turns once clockwise, subtract one from the index. The index is often called the *winding number* of the vector field around the zero.

Consider the sink in figure 19.6. We see the dial in eight locations around the circle. As the dial traverses the circle once counterclockwise, the hand turns once counterclockwise. So the index of a sink is 1. For the saddle, the hand turns once clockwise as the dial follows the circle around the zero once counterclockwise. So the index of a saddle is −1. In a similar manner we can compute the indices of the other zeros in figure 19.4. The sink and the center have index 1, whereas the dipole has index 2.

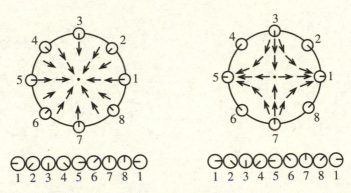

Figure 19.6. A sink has index 1 and a saddle has index −1.

Figure 19.7. The index of a saddle is $2(-1)+1 = -1$ and the index of a sink is $6(-1)+7(1) = 1$.

We now give a second way of computing the index of the zero of a vector field. This approach will be useful to us later. Suppose we put the zero in the interior of a polygonal face (it is acceptable to have curved edges). We are free to chose any polygon as long as a few criteria are satisfied. As before, we must choose the polygon so that our zero is the only zero in the polygon and the polygon must bound a disk. In addition, we insist that every vector that lies on an edge points either inward or outward. We do not want a vector to point along an edge. (Such a polygon always exists, although this is not obvious.) In figure 19.7 we see a saddle placed inside a square and a sink inside a hexagon in such a way that all vectors are either pointing inward or outward.

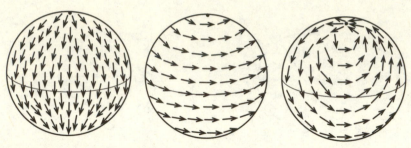

Figure 19.8. Three vector fields on the sphere.

We now single out those edges and vertices on which the vector field is pointing inward. Along each such edge place a −1, and along each such vertex place a 1. Finally, put a 1 in the middle of the polygon. It turns out that the sum of all these numbers is the index of the zero. We see that this is true for the saddle and the sink in figure 19.7.

At last we are able to state the Poincaré-Hopf theorem. It gives us a topological way to determine whether a vector field has a zero (or equivalently, a flow has a fixed point). It also gives us insight into the relative numbers of each type of zero on a particular surface.

POINCARÉ-HOPF THEOREM

For any vector field on a closed surface S with only finitely many zeros, the sum of the indices of all zeros equals the Euler number of the surface, $\chi(S)$.

Before proving this theorem, we will give some examples. In figure 19.8 we see three different vector fields on the sphere. The first, the gradient vector field, has a sink and a source (each with index 1), the second has two centers (each with index 1), and the third has one dipole (with index 2). In all three cases the indices sum to 2, the Euler number of the sphere.

Earlier we saw that the gradient vector field on the torus (figure 19.3) has four zeros—one source, two saddles, and one sink. The sum of their indices is $1 + 2(-1) + 1 = 0$, the Euler number of the torus.

As an added bonus, gradient vector fields enable us to compute the Euler number of surfaces without drawing vertices, edges, and faces. In figure 19.9 see the sphere placed in a bent U-shape. The gradient vector field has two sources, one saddle, and one sink; thus the sum of the indices is $2(1) + 1(-1) + 1(1) = 2$. The double torus has one source, four saddles,

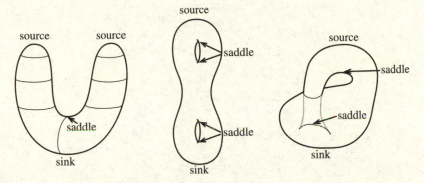

Figure 19.9. The Euler numbers of a sphere, a double torus, and a Klein bottle are 2, −2, and 0, respectively.

and one sink, so χ(double torus) $= 1 + 4(-1) + 1 = -2$. The Klein bottle has a source, two saddles, and a sink, so χ(Klein bottle) $= 1 + 2(-1) + 1 = 0$. We summarize this situation as follows:

> If the zeros of the gradient vector field of a surface S consist of sources, saddles, and sinks, then $\chi(S) = (\#\text{sources}) - (\#\text{saddles}) + (\#\text{sinks})$.

The Poincaré-Hopf theorem states that if a vector field on a surface has finitely many zeros, then the sum of the indices is the Euler number. From this we conclude that if a vector field has no zeros, then the Euler number of the surface must be zero. So any vector field on a surface with nonzero Euler number must have at least one zero! The Euler number of the sphere is 2; so every vector field on a sphere must have a zero. This famous theorem, first proved by L. E. J. "Bertus" Brouwer (1881–1966) in 1911,[2] has the memorable name *the hairy ball theorem*. It is so-named because if we think of a hairy ball (a tennis ball or a hedgehog) as a sphere with a vector field, then it is impossible to comb it without introducing a cowlick. It is also said that "you can't comb the hair on a coconut."

> HAIRY BALL THEOREM
> Every vector field on a sphere has a zero.

From this theorem we can draw the conclusion mentioned in the introduction: that there is always a location on earth where there is no wind. If we view the earth as a sphere, then the surface winds yield a vector

Figure 19.10. Wind vectors on the surface of the earth.

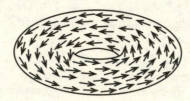

Figure 19.11. A vector field with no zeros on the torus.

field. By the hairy ball theorem, there is a spot on earth where that vector field is zero. The example shown in figure 19.10 has a point of no wind in the center of a cyclone off the coast of South America. (Actually, because this zero has index 1, there must be at least one more windless spot on the other side of the earth!)

Because the Euler number of the torus is zero, the Poincaré-Hopf theorem does not guarantee that every vector field has a zero. Indeed, in figure 19.11 we see an example of a nonvanishing vector field on the torus.

The hairy ball theorem is an example of an "existence theorem." They are ubiquitous in theoretical mathematics. They are simultaneously very powerful and frustratingly imprecise. On the one hand, given only a simple set of hypotheses (a vector field on a sphere), we can make the confident assertion of the existence of some object (a zero vector). However, it is often the case that neither the statement nor the proof of an existence theorem gives a technique for finding this object. We know there is another windless location on the other side of the globe in figure 19.10, but it could

Figure 19.12. Henri Poincaré.

be anywhere, and moreover, there may be one or there may be many. It is like searching for a child's misplaced teddy bear at bedtime—you know it is in the house, but it could be anywhere—under the bed, in the closet, or in the microwave oven. Although more techniques need to be employed to find these objects, often the existence of the object is sufficient for the purpose at hand.

The Poincaré-Hopf theorem is named after the two mathematicians who contributed most to its development, although several others made important contributions as well.

Henri Poincaré was born in Nancy, France, in 1854 into a well-respected upper-middle-class family (his first cousin Raymond Poincaré would later become the president of the French Republic).

Henri's mathematical talent became clear early, with one of his teachers calling him a "monster of mathematics."[3] He began making important contributions to mathematics while still in his twenties, and was elected to the *Académie des Sciences* at the age of thirty-three. He was the stereotypical mathematical genius; he was clumsy, had poor vision, was absent minded. He also possessed a supreme intellect and had the ability to retain and juggle many abstract concepts in his head.

Poincaré is generally acknowledged as the preeminent mathematician of his era. He was the last great universalist. Like Euler and Gauss, Poincaré was an expert in nearly every area of mathematics, pure and applied. He was a voracious reader who was familiar with the latest results. Also like Euler (but unlike Gauss), Poincaré published extensively. He penned nearly five hundred articles, as well as many books and sets of lecture notes. He made important and lasting contributions in such diverse fields as function theory, algebraic geometry, number theory, ordinary and partial differential equations, celestial mechanics, dynamical systems, and, of course, topology. He also published many articles in theoretical physics. Poincaré had a restless curiosity that kept him moving from topic to topic. He would attack a new area of mathematics, make an indelible mark, then move on to the next. A contemporary called him "a conqueror, not a colonist."

It is remarkable that not only was he able to produce the highest-level mathematics, he was also able to write at a level accessible to nonspecialists. He wrote numerous lucid and fascinating publications about science and mathematics for a lay audience. His writings were widely read and translated into many languages.

Poincaré's expertise spanned all of mathematics, but throughout his career he kept returning to the study of differential equations. His successes in this area were profound. According to the mathematician Jean Dieudonné (1906–1992), "The most extraordinary production of Poincaré . . . is the qualitative theory of differential equations. It is one of the few examples of a mathematical theory that sprang apparently from nowhere and that almost immediately reached perfection in the hands of its creator."[4] The prime example was his discovery of the index formula.

His first contribution was in 1881. In this work, he took a differential equation and produced a vector field on the sphere. He defined the index of a zero and proved that the sum of the indices of all zeros is 2.[5] Of course, it is no coincidence that the sum equals 2, for that is the Euler number of the sphere. Poincaré made this observation precise in 1885 when he proved that the sum of the indices of a vector field on a surface is the Euler number of the surface.[6] The following year he defined an index for zeros of a vector field in n-dimensional space, and he sketched the idea of an n-dimensional index theorem. The difficulty in following through with this program was that the topological machinery did not yet exist (this, as we will see in chapter 23, Poincaré would later create).

In 1911 Brouwer properly generalized Poincaré's index theorem to the n-dimensional sphere, S^n. We are familiar with S^1, the unit circle in the plane ($x^2 + y^2 = 1$), and S^2, the unit sphere in 3-dimensional space ($x^2 + y^2 + z^2 = 1$). More generally, S^n is the collection of points 1 unit away from the origin in ($n+1$)-dimensional space ($x_1^2 + x_2^2 + \cdots + x_{n+1}^2 = 1$). Brouwer proved that for any vector field on S^n, the sum of the indices of the zeros is 0 if n is odd and 2 if n is even.[7] In chapters 22 and 23 we will discuss the

Figure 19.13. Heinz Hopf.

Euler number in higher-dimensional spaces. When we do, we will discover that $\chi(S^n) = 0$ when n is odd and $\chi(S^n) = 2$ when n is even.

The next major contributor was Heinz Hopf (1894–1971). Hopf was born in Breslau, Germany (now Wroclaw, Poland). His work in topology made a profound impact on twentieth-century mathematics. A student of Hopf's wrote, "Hopf selected deep problems with an unerring instinct and let them mature. Then he presented in one piece a solution that showed new thoughts and methods."[8]

In his memoirs Hopf pinpoints the pivotal moment in his mathematical career as a two-week period in 1917 when he was on leave from his military service during World War I. He sat in on a mathematics class at the University of Breslau during a presentation of a topological theorem of Brouwer's. After serving on the Western Front, being wounded twice, and receiving the Iron Cross, he resumed his study of mathematics at the University of Breslau. His mathematical career took him to several German universities, Princeton University, and finally to Eidgenössische Technische Hochshule (ETH) in Zurich, Switzerland.

Two years after he arrived in Switzerland, the Nazi party came to power in Germany. Although he was raised Protestant, his father was born Jewish. Solomon Lefschetz and others at Princeton urged Hopf to return, but he and his wife refused to leave Switzerland, and instead worked to help refugees from Germany. Eventually the German government threatened to revoke his citizenship if he did not return. Reluctantly, he gave up his German citizenship and became Swiss. After the war Hopf remained in Switzerland and worked diligently to reestablish mathematics in Germany.

Of Hopf's many important contributions to topology, some of the first were on the topology of vector fields. Beginning in 1925 he published a series of papers generalizing Poincaré's index theorem.[9] We stated the Poincaré-Hopf theorem for surfaces, but Hopf proved that it applies to higher-dimensional generalizations of surfaces, called manifolds (we will learn more about manifolds in chapter 22).

Although the Poincaré-Hopf theorem is usually stated for closed surfaces, mathematicians discovered various generalizations of the theorem. There is an extremely general version for surfaces with boundary,[10] but we will state the following simpler version.

POINCARÉ-HOPF THEOREM FOR SURFACES WITH BOUNDARY

Suppose a surface with boundary S has a vector field with finitely many zeros. If the vector field points inward on every boundary component (or outward on every boundary component), then the sum of the indices of all zeros is the Euler number of the surface, $\chi(S)$.

The hairy ball theorem does not apply to hair on our heads because the hair-covered region of our head is not topologically a sphere—it is a disk. Indeed, the "slicked-back" and "ponytail" styles have no cowlick. However, on a "crew cut" head we see the roots of a person's hair often grow away from the center of the head—downward in the back, toward the ears on the sides, and into the face from the front. Because this hairy vector field points outward along the boundary, the sum of the indices of the zeros must equal $\chi(\text{disk}) = 1$. So there must be a cowlick. The author's baby daughter (whose hair is still just peach fuzz) has the beginnings of three cowlicks, two outward spirals (each of index 1) with a saddle between them (index -1).

We now sketch a proof of the Poincaré-Hopf theorem for surfaces without boundary. (It is not difficult to modify this proof to handle the surfaces with boundary.) The proof is based on one by William Thurston (b. 1946).[11]

We begin with a carefully chosen partition of the surface. First, put each zero of the vector field inside its own polygonal face. These faces can have any shape, with any number of sides, as long as no vector that lies on an

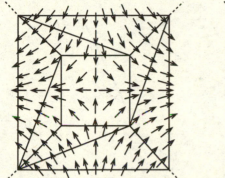
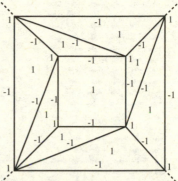

Figure 19.14. A partition of a surface with at most one zero in each face, and the vertices, edges, and faces labeled accordingly.

edge is parallel to the edge. That is, every vector on the boundary of the face must point in or out.

At this point we have faces enclosing all of the zeros of the vector field. Complete the partition by triangulating the rest of the surface. We can do this in any way that we want, although, just as with the earlier faces, we insist that all vectors on the boundaries of the triangles point inward or outward, not along the edge (see figure 19.14).

Now, place a 1 on each vertex, a −1 on each edge, and a 1 in the middle of each face. So when we add all of these quantities over the entire surface, we will obtain $V - E + F$, or in other words, $\chi(S)$. More specifically, because each edge bounds two faces and the vector field points into one of these two faces, put an edge's −1 in the face into which the vectors point. Similarly, each vertex is located at the junction of several faces, but there is one face into which the vector points. Put the 1 in this face (see figure 19.14).

First we examine the triangular faces that do not contain zeros of the vector field. As we see in figure 19.15, there are only two possible situations. One is that the vectors point inward on one edge and no vertices. The other is that the vectors point inward on two edges and on the vertex between them. In either case, the sum of the 1s and −1s is zero. Hence, these triangular faces contribute nothing to the Euler number.

On the other hand, recall that for the faces containing a zero, this technique can be used to compute the index. So each face that contains a zero contributes to the sum a value equal to the index of the zero. As claimed by the Poincaré-Hopf theorem, the sum of all the 1s and −1s is equal to both the Euler number and to the sum of the indices of the zeros.

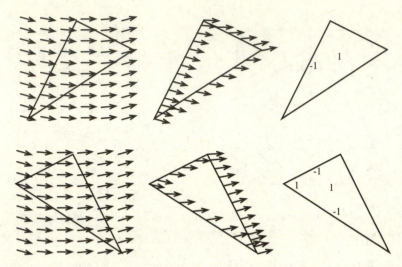

Figure 19.15. The triangles containing no zeros contribute nothing to the Euler number.

$f(y)=y$ x $f(x)$

Figure 19.16. The point y is a fixed point for f and x is not.

As stated, the Poincaré-Hopf theorem is a theorem about vector fields, but because vector fields can be used to construct flows, it can also be interpreted as a fixed point theorem for continuous dynamical systems. We conclude this chapter by mentioning another famous fixed point theorem.

A flow on a surface is a mathematical way of describing the continuous movement of particles. We now look at a related, but quite different situation. Suppose that instead of flowing, each point in a surface S jumps to a new location. Mathematically, we describe this motion using a continuous function f with domain S and range S (by continuous we mean simply that nearby points jump to nearby points). A point begins at a point x, then it moves to a new point $f(x)$. Just as for flows, we are especially interested in those points that do not move. A point y in S is called a *fixed point* for f if $f(y) = y$ (see figure 19.16).

Probably the most famous fixed point theorem is the *Brouwer fixed point theorem*. It applies to continuous functions from the n-ball to itself.

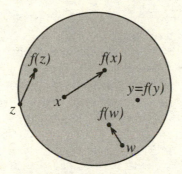

Figure 19.17. The vector field for a function from B^2 to itself.

The n-ball, denoted B^n, is the set of all points in n-dimensional space that are within one unit of the origin. That is, it is the set of points satisfying $x_1^2 + x_2^2 + \cdots + x_{n+1}^2 \leq 1$, or simply, B^n is the set of points contained inside the $(n-1)$-dimensional sphere, S^{n-1}. Brouwer proved the result for B^3 in 1909[12] and for B^n ($n > 3$) in 1912.[13]

BROUWER FIXED POINT THEOREM
Any continuous function from B^n to itself must have a fixed point.

Here is one way to think about this remarkable theorem. Consider the case $n = 2$. B^2 is a disk in the plane—the region enclosed by S^1, the unit circle. Imagine that it is a dinner plate. Cover the plate with a piece of paper that is at least as large as the plate, then cut the part of the paper that hangs off the edge of the plate. Now, pick up the paper, crush it into a ball (being careful not to rip it), and place it back on the plate. The Brouwer fixed point theorem says that some part of the paper is over exactly the same spot that it was originally. This same reasoning says that a designer of a one-story shopping mall can place a map of the mall anywhere in the building and be able to put a star on the map labeled "you are here."

The proof of this theorem follows easily from the Poincaré-Hopf theorem (the version for surfaces with boundary). We will consider the case when $n = 2$, but the argument for larger n is identical. Begin with a function f from B^2 to itself. Define a vector field on B^2 as follows: to each point x in B^2 assign the vector that has its tail at x and tip at $f(x)$ (see figure 19.17). B^2 is a surface with boundary and all vectors on the boundary point inward, so we may apply the Poincaré-Hopf theorem. Because $\chi(B^2) = 1 \neq 0$, the vector field must have at least one zero. In this case, a zero vector corresponds to a point y for which $f(y) = y$. In other words, f must have at least one fixed point.

In fact, the Brouwer fixed point theorem applies equally well to any shape that is homeomorphic to B^n. The coffee in a coffee cup is homeomorphic to B^3. Swirl the coffee vigorously in the cup (without spilling it!), and according to the Brouwer fixed point theorem, after it comes to rest, there is a molecule of coffee in precisely the same location as it was at the start.

In this chapter we observed that the topology of an object, measured simply by its Euler number, can force global behavior that does not appear to have any relationship to the global topology—the existence of fixed points of flows or functions. As we will see in the next two chapters, the topology of a shape can also determine certain of its global geometric properties.

CHAPTER 20

WHEN TOPOLOGY CONTROLS GEOMETRY

And now remains,
That we find out the cause of this effect,
Or rather say the cause of this defect,
For this effect defective comes by cause.
Thus it remains and the remainder thus.
—William Shakespeare, Hamlet[1]

For most of this book we have been moving away from the rigid confines of geometry, working instead in the much more fluid environment of topology. In this chapter and the next we return to geometry. We will examine polygons, polyhedra, curves, and surfaces, made not of rubber, but of the hardest steel. However, these geometric objects can still be viewed with a topological eye—the polygons and curves are homeomorphic to a circle, and the polyhedra and surfaces are homeomorphic to a sphere or a g-holed torus.

We will present a collection of theorems that shows the surprising relationship between the topology of these shapes and their geometry. We will see how the Euler number can predict certain geometric properties. Our ultimate goal is to present three theorems. In this chapter we will see Descartes' formula for polyhedra and the angle excess theorem for surfaces, and in the next chapter we will investigate the Gauss-Bonnet theorem for surfaces. These show that certain global geometric properties (related to angles and curvature) are completely determined by topology (given by the Euler number). In this way we will see how topology can control geometry.

Before looking at these theorems for polyhedra and surfaces, we will investigate the analogous results in dimension one. The 1-dimensional analogues of a polyhedron and a surface are a polygon and a simple closed curve, respectively. The first theorem is encountered in any elementary geometry class (see figure 20.1).

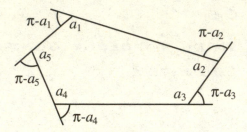

Figure 20.1. The exterior angles of a polygon.

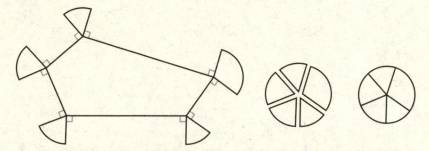

Figure 20.2. The exterior angles of a polygon sum to 2π.

EXTERIOR ANGLE THEOREM
The sum of the exterior angles of a polygon is 2π.

George Pólya (1887–1985) discovered the following short and elegant proof of the exterior angle theorem for convex polygons.[2] At each corner, draw two line segments pointing outward, one line segment perpendicular to each side (see figure 20.2). Draw a sector of a unit circle at each vertex with these segments as sides. Observe that the angle made by these two segments is precisely the exterior angle. This is true because the two right angles add up to π, so the interior angle and the angle of the sector must also add up to π. Because the sides of each pair of adjacent sectors are parallel, we can reassemble the sectors to form a circle. So, the sum of the exterior angles is 2π. We omit the proof for nonconvex polygons, but it follows from the observation that any nonconvex polygon can be decomposed into convex ones.

In a sense, the exterior angle theorem is not surprising. A car driving along a polygonal road would have to turn at each corner, and the size of each turn is the exterior angle. In order to return to the starting position, the car would have to turn through a total angle of 360°.

Although the typical adult may be hard pressed to remember the quadratic formula or the Pythagorean theorem, there is one mathematical result that almost any adult can quote: the sum of the interior angles of a triangle is 180° (or as we say, π radians). The 180° theorem is a simple consequence of the exterior angle theorem. If a, b, and c are the interior angles of a triangle, then $\pi - a$, $\pi - b$, and $\pi - c$ are the exterior angles. By the exterior angle theorem $(\pi - a) + (\pi - b) + (\pi - c) = 2\pi$. Rearranging terms yields $a + b + c = \pi$.

For polygons other than triangles, the interior angles sum to more than 180°, but the sum still depends only on the number of sides. If $a_1, \ldots, a_n$ are the interior angles of a polygon, then by the exterior angle theorem

$$2\pi = (\pi - a_1) + (\pi - a_2) + \cdots + (\pi - a_n).$$

Rearranging terms yields the following useful theorem.

INTERIOR ANGLE THEOREM
The sum of the interior angles of a polygon with n sides is $(n - 2)\pi$.

To ease the transition to Descartes' formula for polyhedra, it will be helpful to look at the exterior angles of a polygon in a slightly different way. Think of the corners of a polygon as being "imperfect" straight lines. Looking at each angle, we ask how much the bend differs from a straight line. At a corner with interior angle a the curve differs from a straight line by $\pi - a$, the exterior angle. Taking this point of view, we will call $\pi - a$ the *angle deficit* or *angle defect* of the corner. So we rephrase the exterior angle theorem as follows.

EXTERIOR ANGLE THEOREM (REPHRASED)
The total angle deficit of any polygon is 2π.

There is a smooth analogue of the exterior angle theorem. Consider, again, the car analogy. A Grand Prix race course is a winding, curvy road that loops back to its starting point. As a Formula One race car navigates the track it swerves right, left, right, and left, but by the time it returns to the starting line it has made one complete circuit counterclockwise. In other words, allowing the left and right turns to cancel, the car winds through a total 360° of left-hand turns.

 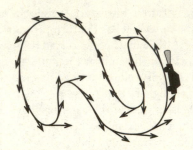

Figure 20.3. The tangent vectors of simple closed curves turn through 2π.

Now, consider a smooth simple closed curve in the plane (the race course; see figure 20.3). Pick an orientation on the curve, then place tangent vectors on the curve pointing in this direction (the headlights of the car). We are interested in the behavior of these tangent vectors as we proceed once around the curve. If the curve is a circle, then as we go around the circle one time counterclockwise, the vectors turn around one time counterclockwise as well—they turn through an angle of 2π. Here it may be helpful to think of a tangent vector as being the needle of a dial. As we move the dial around the circle, the needle turns around exactly one time counterclockwise. For a more complicated curve, as the dial traverses the curve, the needle may move forward and backward, but in the end it travels exactly once around the dial.

Although this observation may seem obvious (as it was believed to be for a long time), it is difficult to prove. In 1935 Hopf proved the theorem.[3] It is known as *Hopf's Umlaufsatz*, or more simply, the *theorem of turning tangents*.

Theorem of Turning Tangents

The tangent vectors on a smooth simple closed curve in the plane turn through an angle of 2π.

It is not difficult to see the relationship between the exterior angle theorem and the theorem of turning tangents. In fact, it is possible to state a theorem that is a hybrid of these two in which the curve is smooth except for a finite number of sharp corners. A car driving on a curvy road that occasionally has to make sharp turns, turns through 360° by the time it returns to its starting location.

Returning to the original assertion, we ask, how do these theorems bridge two mathematical subjects? They show a sense in which topology

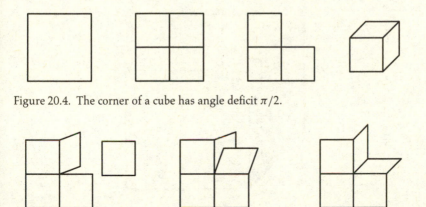

Figure 20.4. The corner of a cube has angle deficit $\pi/2$.

Figure 20.5. This corner has angle excess $\pi/2$.

can control geometry. A topologist cannot tell the difference between polygons and smooth simple closed curves. They are all circles. A topologist does not talk about angles, straightness, tangent vectors, and so on. To a geometer every polygon and every simple closed curve is different from the rest, and he describes these objects by speaking of their corners, their curvature, and other descriptors. The exterior angle theorem and the theorem of turning tangents say that being homeomorphic to a circle completely determines a geometric property—the total angle deficit of the shape. Regardless of how many bends it has, the total angle deficit is 2π.

We now investigate how these two theorems generalize to Descartes' formula for polyhedra and the angle excess theorem for surfaces.

Get a square piece of paper, a pair of scissors, and tape. Divide the paper into four equal quadrants, and use the scissors to cut one of them away (set this piece aside for later). Then tape together the two cut edges to obtain an object that looks like the corner of a rectangular box (see figure 20.4).

We defined the angle deficit at a corner of a polygon to be the amount by which the bend failed to be a straight line. Likewise, define the angle deficit of a solid angle to be the amount by which it fails to be a flat plane. In our example, four right angles (2π) met at the center of our paper, then we cut one away (leaving $3\pi/2$). Thus the angle deficit at the corner of a cube is $2\pi - 3\pi/2 = \pi/2$.

Take another square piece of paper. As before, divide it into quadrants. Cut once from the edge to the center (see figure 20.5). Take the discarded piece from the first construction and tape the two cut edges to the cut on

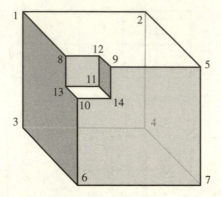

Figure 20.6. This nonconvex polyhedron still has total angle deficit 4π.

the folded paper. Doing so we see that there is too much angle present. We get a configuration that resembles a brick wall with a brick removed. The central vertex has a total angle of $5\pi/2$, and that is $\pi/2$ more than a flat plane. We call this an angle deficit of $-\pi/2$, or an *angle excess* of $\pi/2$.

A polyhedron has many vertices, each with its own angle deficit (or angle excess). The *total angle deficit* of a polyhedron is the sum of all of its angle deficits.

Consider some examples. Each of the eight corners of a cube has an angle deficit of $\pi/2$, so the total angle deficit is 4π. The four faces of a tetrahedron are equilateral triangles. Since three equilateral triangles meet at each vertex, the angle deficit at each corner is $2\pi - 3(\pi/3) = \pi$. There are four vertices, so the total angle deficit is 4π. Finally, consider the nonconvex polyhedron in figure 20.6: a large cube with a smaller cube removed from one of its corners (think of a Rubik's cube with a corner piece pulled out). The corners marked 1–10 have angle deficit $\pi/2$. Corner 11 is facing the "wrong way" but it still has angle deficit $\pi/2$. The three remaining corners (12, 13, and 14) have angle excesses of $\pi/2$. Thus the total angle deficit is $11(\pi/2) + 3(-\pi/2) = 4\pi$.

At this point the pattern is becoming clear, and we conjecture that every polyhedron has a total angle deficit of 4π. This observation was first made by Descartes in his unpublished notes that we discussed in chapter 9, *The Elements of Solids*. The third sentence of Descartes' notes reads:

> As in a plane figure [polygon] all the exterior angles, taken together, equal four right angles [2π], so in a solid body [polyhedron] all the exterior solid angles [angle deficits], taken together, equal eight solid right angles [4π].[4]

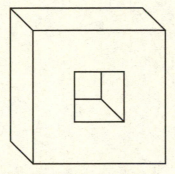

Figure 20.7. The total angle deficit of the torus is zero.

As Descartes pointed out, the parallels with the exterior angle theorem are self-evident. Just as the sum of the angle deficits of a polygon is 2π, the sum of the angle deficits of a polyhedron is 4π.

A slight variant of this theorem was rediscovered by Euler and appeared in his papers on the polyhedron formula.[5] Euler proved that the sum of all plane angles in a polyhedron with V vertices equals $2\pi(V-2)$. Where Descartes' formula generalizes the exterior angle theorem for polygons, Euler's formula generalizes the interior angle theorem. It is easy to see that Euler's result is equivalent to Descartes'. The total angle deficit is simply $2\pi V$ minus the sum of all plane angles, or $2\pi V - 2\pi(V-2) = 4\pi$.

Of course, both Euler and Descartes were considering convex polyhedra. It turns out that with a little modification, the theorem applies to all polyhedra, even those that are not topological spheres. The total angle deficit is a topological invariant, and it has a simple relationship with the Euler number of the polyhedron.

DESCARTES' FORMULA
The total angle deficit of any polyhedron P is $2\pi\chi(P)$.

The cube, tetrahedron, and broken cube are topological spheres and have Euler number 2, so the total angle deficit is $2\pi\chi(P) = 2\pi \cdot 2 = 4\pi$. As a nonspherical example, consider the polyhedral torus shown in figure 20.7. It has sixteen corners, eight of which have an angle deficit of $\pi/2$ and eight which have an angle excess of $\pi/2$ (angle deficit of $-\pi/2$). So, the total angle deficit is zero, the Euler number of the torus. The reader is encouraged to verify Descartes' formula for the paper polyhedra in appendix A.

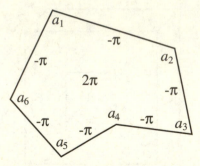

Figure 20.8. For an n-gon, $(a_1 + \cdots + a_n) - n\pi + 2\pi = 0$.

We now prove Descartes' formula. Let P be a polyhedron with V vertices, E edges, and F faces, and let T be the total angle deficit of P. We must show that $T = 2\pi \chi(P) = 2\pi V - 2\pi E + 2\pi F$.

Choose any face of the polyhedron. Suppose it has plane angles $a_1, \ldots, a_n$. By the interior angle theorem,

$$a_1 + \cdots + a_n = (n - 2)\pi.$$

Rearranging the terms yields

$$(a_1 + \cdots + a_n) - n\pi + 2\pi = 0.$$

We can visualize this equality as follows. If we put $-\pi$ on each edge of the face, the angle measure at each corner, and 2π in middle of the face (see figure 20.8), then the sum of these quantities is zero.

Now, do the same for every face of P and add up all of these quantities. Each face contributes 2π and each edge contributes -2π ($-\pi$ from each side). So,

$$S - 2\pi E + 2\pi F = 0,$$

where the S is the sum of all plane angles in P. Now add T, the total angle deficit, to each side of the equality,

$$(T + S) - 2\pi E + 2\pi F = T.$$

Because T is the total angle deficit, then by adding T, we are adding enough so that the angle sum at each vertex can again be 2π. In other words, $T + S$ is $2\pi V$. So, $T = 2\pi V - 2\pi E + 2\pi F = 2\pi \chi(P)$.

Descartes' formula is a beautiful illustration of the surprising relationship between topology and geometry. Because the total angle deficit is related to the Euler number, we see that the topology of the polyhedron completely determines one aspect of its global geometry.

As an application of this theorem, the reader is encouraged to find a new proof that there are only five Platonic solids.

In most of this book we have assumed that the edges that partition a surface into faces are topological entities. The edges can wiggle crazily and create faces that are wildly misshapen. In this chapter we are investigating the much less wild and crazy field of geometry. Ideally we would like our faces to be polygons with straight edges. On a curved surface it is impossible for the edges to be straight, so instead we require that they be *geodesic curves*.

In chapter 10 we introduced the concept of a geodesic on a sphere. It was any segment of a great circle. It turns out that we can define a geodesic curve on any rigid surface. It is characterized by minimizing length—the shortest path between two points on a surface is realized by traveling along a geodesic curve. The well known saying, "The shortest distance between two points is a straight line" should be "The shortest distance between two points is a geodesic curve." For the rest of this chapter we assume that the edges on our surfaces are geodesic curves, making the faces *geodesic polygons*.

One benefit of working with geodesic polygons is that we can measure the angles at the vertices. The edges are curved, but when we zoom in to one of the angles with a strong microscope (figuratively speaking), the edges appear straight, so it is possible to measure the angles.

Triangles in the plane obey the 180° theorem, but on a typical surface, the 180° theorem need not apply. Recall that Harriot and Girard proved that the sum of the interior angles of a geodesic triangle on a sphere exceeds 180° (chapter 10). There are other surfaces, such as saddle-shaped surfaces, for which the sum of the interior angles of a geodesic triangle will be less than 180° (see figure 20.9).

So we can speak about the angle excess or angle deficit of a geodesic triangle—the amount that the interior angle sum differs from that of a planar triangle. That is, the *angle excess* of a geodesic triangle with interior angles a, b, and c is $(a+b+c)-\pi$. If $(a+b+c)-\pi$ is negative, the triangle has an *angle deficit*.

Similarly we can define the angle excess or deficit of an n-sided geodesic polygon. By the interior angle theorem, the sum of the interior angles of a planar n-gon is $(n-2)\pi$. So the angle excess of an n-gon with interior angles $a_1, a_2, \ldots, a_n$ is $(a_1 + a_2 + \cdots + a_n) - (n-2)\pi$.

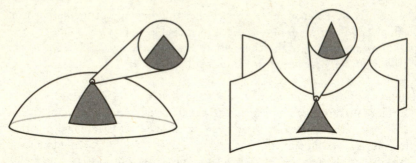

Figure 20.9. A triangle with an angle excess (left) and one with an angle deficit (right).

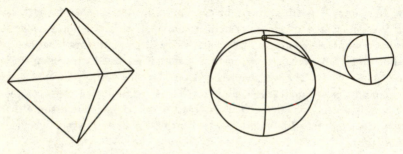

Figure 20.10. An octahedron rolled into a ball.

It is important not to confuse angle excess and deficit for polyhedra and for surfaces. A polyhedron has an angle excess or deficit at its vertices, whereas it is the faces on a surface that have angle excesses or deficits. It is confusing that they have the same names, but as we see in the following discussion, they are intimately related.

Take a lump of modeling clay and make an octahedron. Every face is an equilateral triangle, so the angle deficit at each vertex is $2\pi - 4(\pi/3) = 2\pi/3$. Since there are six vertices, the total angle deficit is $6(2\pi/3) = 4\pi$, as guaranteed by Descartes' formula. Darken each of the edges using a marker. Then place the polyhedron on a table and roll it until it is spherical (figure 20.10). What were triangular faces in the octahedron are now curved. If we roll this shape carefully, the straight edges turn into geodesic segments and the faces become geodesic triangles.

After rolling the octahedron into a ball, there is no longer an angle deficit at any vertex. Rolling it on the table flattens the corners so the angles sum to 2π. Where did the angle deficit go?

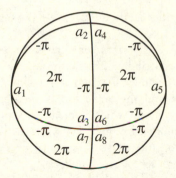

Figure 20.11. Labeling a surface with 2π on each face, $-\pi$ on each side of each edge, and the angle measures at each vertex.

It is not difficult to see that during this process the interior angles of the triangles changed size. The angles meeting at each vertex, which measured 60° before, are now right angles. The triangles on the clay ball have three right angles, so the sum of the interior angles is $3\pi/2$. The triangular faces have an angle excess. The angle deficits at the vertices of the octahedron get distributed among the faces of the ball to become angle excesses in the triangles. Similarly, for any partition of a surface by geodesic polygons, the vertices no longer have angle deficits or angle excesses, but the faces do.

If a surface is partitioned into vertices, geodesic edges, and faces, then the *total angle excess* is the sum of the angle excesses of all faces. Just as the total angle deficit of a polyhedron is related to its Euler number (Descartes' formula), the total angle excess of a surface is related to its Euler number. We have the following analogue of Descartes' formula for surfaces.

ANGLE EXCESS THEOREM FOR SURFACES
The total angle excess of a surface S is $2\pi\chi(S)$.

The proof of this theorem should look familiar by now. Let S be a surface partitioned into vertices, geodesic edges, and faces. Decorate the surface by putting 2π in the middle of each face, $-\pi$ on each side of each edge, and the angle measure at each angle (see figure 20.11). If we sum these quantities on a particular n-sided face with interior angles $a_1, a_2, \ldots, a_n$ we obtain the angle excess of the face,

$$2\pi - n\pi + (a_1 + a_2 + \cdots + a_n) = (a_1 + a_2 + \cdots + a_n) - (n-2)\pi.$$

Thus, adding all quantities on the surface gives the total angle excess of the surface.

On the other hand, each vertex contributes 2π, each edge contributes -2π, and each face contributes 2π. Summing all values yields $2\pi V - 2\pi E + 2\pi F = 2\pi \chi(S)$, and we have our result.

Descartes' formula and the angle excess theorem are two beautiful theorems that show a sense in which topology can control geometry. In the next chapter we see another example. We will see that the total curvature of a surface depends on the surface's topology, and that it depends in an intimate way on the surface's Euler number.

CHAPTER 21

THE TOPOLOGY OF CURVY SURFACES

If others would but reflect on mathematical truths as deeply and as continuously as I have, they would make my discoveries.
—Carl Friedrich Gauss[1]

One of the most fundamental topics in the geometry of planar curves is curvature. The *curvature* at a point x is a number, k, that measures the "sharpness" of the turn at x—it measures how quickly the tangent vectors change direction. Roughly speaking, given a normal vector $\vec{n}$ to a curve at x, if the curve bends in the direction of $\vec{n}$, then $k > 0$, if it bends away, then $k < 0$, otherwise $k = 0$ (see figure 21.1). The more sharply it bends, the larger (in absolute value) k is.

By the Jordan curve theorem a simple closed curve in the plane has an inside and an outside. So we can pick normal vectors along the entire curve, having them all point toward the inside. Then we can compute the curvature at each point on the curve. Typically, the curvature varies from point to point (see figure 21.2). It is possible to sum the curvature over the entire curve to obtain the *total curvature*. The details of this computation are beyond the scope of this book, but a calculus student should recognize that because the curvature varies continuously, the sum in question is the integral of the curvature. We have the following theorem.*

> TOTAL CURVATURE THEOREM FOR CURVES
> The total curvature of any smooth simple closed plane curve is 2π.

In other words, the total curvature is the same number for all smooth simple closed curves! If we took a loop of string and dropped it on a

---

* A mathematician would state the theorem as $\int_C k\, ds = 2\pi$, where C is a smooth simple closed curve.

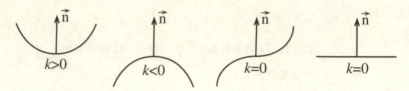

Figure 21.1. Curves with $k > 0$, $k < 0$, $k = 0$, and $k = 0$ (left to right).

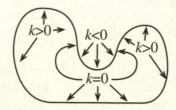

Figure 21.2. A curve with regions of positive, negative, and zero curvature for normal vectors pointing toward the inside.

table so that it did not cross itself, then the regions of negative curvature would cancel with regions of positive curvature, so that in the end, the total curvature is 2π. That is, being homeomorphic to a circle completely determines the total curvature. Again we see how topology controls geometry.

We will not prove this theorem, but it is closely related to the theorem of turning tangents from the previous chapter. Again, a calculus student sees that because we are summing the rate of change of the turning tangents, the total curvature is just the total change in the turning tangents, or 2π.

We can also think of this theorem as another generalization of the exterior angle theorem for polygons. A polygon has no curvature along its sides, instead it carries all of its curvature at the corners in the form of the exterior angles. The total curvature is 2π.

Now we turn our attention from curves to surfaces. Because we are studying the geometric properties of surfaces, we must assume that the surfaces are rigid, and not made of rubber as were our topological surfaces. We assume that they are smooth, with no sharp creases or corners.

Just as we did with curves in the plane, we investigate the curvature of surfaces in 3-dimensional space. Again, pick a normal vector $\vec{n}$ to the surface at a point x. Then look at a plane that passes through the point x and is parallel to $\vec{n}$. The intersection of the surface and the plane is a curve, so

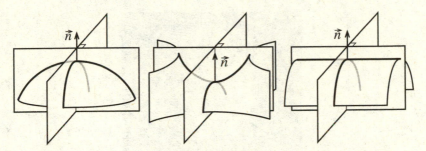

Figure 21.3. Surfaces for which $k_1, k_2 < 0$ (left), $k_1 > 0, k_2 < 0$ (middle), and $k_1 = 0$, $k_2 < 0$ (right).

we can compute its curvature. Typically the curvature of these curves will be different for different planes through x. The largest and smallest values, k_1 and k_2, respectively, are called the *principal curvatures* of the surface at x (see figure 21.3). In 1760 Euler proved that the principal curvatures arise from perpendicular planes.[2]

It was in this way that geometers measured the curvature of surfaces until Gauss made a simple, but crucial modification. He multiplied the principal curvatures to form a single curvature value, now called *Gaussian curvature*: $k = k_1 k_2$. This seemingly trivial operation, which appears to contain less information than do the two principal curvatures separately, produced an important numerical value that helped mathematicians understand the curvature of surfaces.

Surprisingly, most great mathematicians were not child prodigies; their genius needed to grow and mature, only coming out later in life. However, it was clear from a young age that Gauss had tremendous mathematical ability. He was born in 1777 in Brunswick, Germany. At only three-years old Gauss shocked his father, Gerhard, when he pointed out an error in his father's arithmetic in his financial books. On subsequent Saturdays Gauss would sit in a highchair and assist his father.

In his old age Gauss enjoyed telling how, when he was seven-years old, he shocked one of his brutish and overbearing school teachers. The teacher asked the class to sum an arithmetic series $(1 + 2 + 3 + \cdots + 100$, for the sake of the story[3]). Gauss almost immediately wrote the number 5050 on his slate tablet, which he placed on the table in front of the skeptical teacher, and announced "ligget se" (there it lies). Instead of performing the tedious summation, Gauss noticed that by adding the first number to the last, the second to the penultimate, and so forth, each sum would be 101

Figure 21.4. Carl Friedrich Gauss.

(i.e., $1 + 100, 2 + 99, 3 + 98, \ldots$). Because there are fifty such pairs, the total must be $50 \cdot 101 = 5050$.

This classroom incident set in motion a chain of events that in 1791 brought Gauss to the attention of Carl Wilhelm Ferdinand, Duke of Brunswick. The Duke was taken with the fourteen-year-old boy and promised to pay for his schooling. The generous Duke financed Gauss's education at the Collegium Carolinium and University of Göttingen, and continued to pay his salary until the Duke's death at the hands of Napoleon's army in 1807.

The Duke's instinct about the boy was correct. Gauss's first important result, the proof of the quadratic reciprocity law, came when he was nineteen years old. This theorem, which he called *theorema aureum* (the golden theorem), had eluded both Euler and Legendre.

Gauss chose as his seal a tree bearing a few pieces of fruit with the words *pauca sed matura* (few, but ripe). Indeed, this motto perfectly embodies Gauss's career. Unlike the prolific Euler, Gauss was not quick to publish. He never published trivialities, insisting that each publication be a masterpiece. As he said, "You know that I write slowly. This is chiefly because I am never satisfied until I have said as much as possible in a few words, and writing briefly takes far more time than writing at length."[4] Among the many fields upon which Gauss left his mark were astronomy, geodesy, theory of surfaces, conformal mappings, mathematical physics, number theory, probability, topology, differential geometry, and complex analysis.

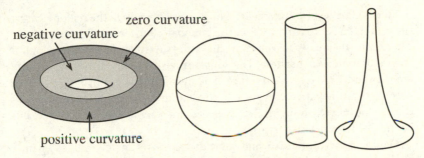

Figure 21.5. A surface (torus) with variable curvatrure: positive, negative, and zero. Other surfaces have constant positive curvature (sphere), zero curvature (cylinder), and constant negative curvature (pseudosphere).

Because of his drive for perfection, he left many great results unpublished. His mathematical diary (*Notizenjournal*), which was discovered forty-three years after his death, contains a wealth of mathematics. Had Gauss published some of these results and nothing else, he would have been remembered as an influential mathematician. It is frustrating to realize that mathematicians toiled for years only to rediscover ideas that had been known to Gauss. One wonders how much further nineteenth century mathematics would have progressed had Gauss been more willing to publish his results.

After the Duke died, Gauss was forced to take a position as the director of the Göttingen Observatory. He spent much of the last twenty years of his life dealing with astronomical affairs at the observatory. He lived to the age of 78, passing away peacefully on February 23, 1855.

Using Gauss's single-valued approach to measuring curvature, $k = k_1 k_2$, we can say that the curvature at a point is positive, negative, or zero. Returning to figure 21.3, we see that if like the bottom of a bowl, both curves bend toward (or both curve away from) the normal vector, then the signs of k_1 and k_2 are the same, and we have a positive curvature. On the other hand, if like a saddle, one curve bends toward the normal vector and one away, the signs of k_1 and k_2 are opposite, and the curvature is negative. If one or both of the principal curvatures are zero, as with a cylinder or a plane, the curvature is zero.

It is important to emphasize that curvature is measured at a single point. A typical surface will have regions of positive, negative, and zero curvature. For instance, the torus in figure 21.5 has positive curvature in the region farthest from the center, negative curvature in the region nearest the center, and zero curvature on the boundary between these two regions. There are surfaces with constant curvature. A sphere (not a topological sphere, but an actual sphere) has constant positive curvature, and a plane and a cylinder have zero curvature. The most famous example of a surface with constant negative curvature is the trumpet-shaped surface known as the *pseudosphere*, so-named not because of a resemblance to a sphere, but because of its constant curvature.

Gaussian curvature, area, and angle excess are intimately related quantities, and it is their relationship that we must understand. We have already seen that curvature and angle excess are related. Figure 20.9 shows a geodesic triangle on a sphere with an angle excess and one on a saddle with an angle deficit. The flatter a surface, the more it resembles a plane, and the more a triangle looks like a planar triangle. Positive curvature yields an angle excess and negative curvature causes an angle deficit.

It should also be clear that size matters. Triangles that are very small experience very little of the surface's curvature (think of two equilateral triangles on the earth, one which has sides 1,000 miles long and one which has sides 1 inch long). As we zoom in to a surface, it looks more and more flat. The smaller the triangle, the closer the angle excess is to zero.

Here is another illustration that curvature and area are related. Suppose we took a portion of a positively curved surface, like a piece of onion skin or a cabbage leaf. If we tried to flatten it out on a table, we would find that there is too much area in the middle. Undoubtedly, the outside edge of the onion skin would rip as we tried to press it flat. This is why in the usual (Mercator) projection of the earth, Greenland dwarfs the continental United States, when in fact three islands the size of Greenland could fit comfortably inside the lower forty-eight states. We run into the opposite problem for negatively curved surfaces. If we were to cut a portion from a saddle-shaped surface there would be too much area at the extremes to flatten the disk onto a table. The outside edge of the disk would fold and wrinkle.

We can exploit the relationship among curvature, area, and angle excess to obtain an alternate definition of Gaussian curvature. Consider a geodesic triangle $\triangle$ containing the point x, with interior angles a, b, and c. The angle excess of this triangle, $E(\triangle) = a + b + c - \pi$, is a good measure of curvature at x. The problem is, as we have already noticed, as the triangle gets smaller, $E(\triangle)$ gets closer to zero. This suggests that we should scale the angle excess by the area. Instead of working with the quantity $E(\triangle)$ we use $E(\triangle)/A(\triangle)$, where $A(\triangle)$ is the area of the triangle $\triangle$. It turns out that as we let $\triangle$ shrink down to x, the quantity $E(\triangle)/A(\triangle)$ approaches the Gaussian curvature at x.

In terms of this formulation, Gaussian curvature is particularly easy to compute for surfaces with constant curvature. Because the curvature is constant, it is simply $E(\triangle)/A(\triangle)$ where $\triangle$ is any geodesic triangle (we do not have to shrink the triangle to a point). For example, take $\triangle$ to be an octant of a sphere of radius r. Such a triangle is made of three right angles, so it has angle excess $E(\triangle) = 3(\pi/2) - \pi = \pi/2$ and area $A(\triangle) = (1/8)4\pi r^2 = \pi r^2/2$. So at every point on the sphere the Gaussian curvature is $(\pi/2)/(\pi r^2/2) = 1/r^2$. This shows that as the radius of the sphere increases, the curvature decreases. It is easy to see the curvature of a billiard ball, but not easy to detect the curvature of the earth.

There is another interesting conclusion that we can draw from this definition of Gaussian curvature. Consider a sheet of paper sitting on a table top. Clearly it has zero Gaussian curvature. If we roll it into a cylinder, the geometry has changed, but the Gaussian curvature is still zero. Try as we might, we will never be able to turn the paper into a positively curved sphere or a negatively curved saddle. The sheet of paper will have zero curvature regardless of how we deform it. In more technical lingo, we are free to alter the *extrinsic curvature* of the paper, but we will never be able to change its *intrinsic curvature*.

The two principal curvatures k_1 and k_2 measure the extrinsic curvature of a surface—they depend on the way the surface is placed in 3-dimensional space. A flat piece of paper has $k_1 = k_2 = 0$, but for a cylinder either k_1 or k_2 is nonzero. The principal curvatures are extrinsic because inhabitants of the surface cannot compute them by making measurements on the surface. They have to leave the surface and examine how it is situated in the ambient space. Because the Gaussian curvature is the product of the principal curvatures, $k = k_1 k_2$, it too is a measure of the extrinsic curvature.

However, area and angle measures are intrinsic properties of a surface, for they can be measured by an inhabitant on the surface. To compute these values, we do not have to place the surface rigidly in space. The area and angle measures of a triangle drawn on the sheet of paper do not change after we roll the paper into a cylinder. So, because we can define Gaussian curvature in terms of these quantities, it actually measures the intrinsic curvature of the surface!

It was Gauss who first discovered that the product of the two extrinsic principal curvatures gave a measure of the intrinsic curvature of the surface. He recognized the beauty of this discovery, so he gave it the grand name of *theorema egregium*, or the excellent theorem.

Because Gaussian curvature is intrinsic, it is not such a rigid measure of curvature that the object must sit immobile in space. However, it is not a topological measurement. If our piece of paper was just a topological surface (made out of rubber), then we could dramatically alter the curvature and greatly distort a triangle drawn upon it.

In 1827 Gauss proved another important theorem that further exploited the relationship among curvature, area, and angle excess.[5] Just as we computed the total curvature of a simple closed curve, Gauss wanted to compute the total curvature of a region of a surface. For a surface of constant curvature, the computation is simple. If the Gaussian curvature is k, then the total curvature for a region R is $k \cdot A(R)$, where $A(R)$ is the area of R. When the region is a geodesic triangle $\triangle$, the total curvature is $k \cdot A(\triangle) = [E(\triangle)/A(\triangle)]A(\triangle) = E(\triangle)$, the angle excess of the triangle.

Gauss's wonderful theorem states that this is true even for geodesic triangles on surfaces of nonconstant curvature.[†]

THE LOCAL GAUSS-BONNET THEOREM
The total curvature of a geodesic triangle on a surface is precisely the angle excess of the triangle.

In other words, this theorem states that total curvature of a geodesic triangle $\triangle$ is $a + b + c - \pi$, where a, b, and c are the interior angles of $\triangle$.

The other individual whose name is associated with this theorem is the French geometer Pierre Ossian Bonnet (1819–1892). In 1848 Bonnet extended Gauss's theorem by proving a version for regions without geodesic sides, which we will not state here.[6] Thus, Gauss was able to compute the total curvature of any geodesic triangle, and Bonnet was able to compute the total curvature of any enclosed region of a surface.

Surprisingly, neither Gauss nor Bonnet asked what seems now to be a natural question: what is the total curvature of the entire surface? They did not even ask about the total curvature of a sphere. It is trivial to compute the total curvature of a surface by combining the local Gauss-Bonnet and angle excess theorems (for technical reasons we must require the surfaces to be orientable).

[†] The local Gauss-Bonnet theorem states that $\int_{\triangle} k\, dA = a + b + c - \pi$, where a, b, and c are the interior angles of the geodesic triangle $\triangle$.

Partition a surface into geodesic triangles. By the local Gauss-Bonnet theorem, the total curvature of each triangle is its angle excess. Thus the total curvature for a surface S is the total angle excess of the surface, which we know is $2\pi\chi(S)$. This is what we now call the *global Gauss-Bonnet theorem*.[‡]

> ### THE GLOBAL GAUSS-BONNET THEOREM
> The total curvature of an orientable surface S is $2\pi\chi(S)$.

Roughly speaking, the global Gauss-Bonnet theorem states if we take a surface and stretch and pull it, we may change the local curvature, but the net curvature does not change. Any new regions of positive curvature will be counteracted by new regions of negative curvature. All that matters is the topology of the surface.

It might seem troubling that the point-wise curvature of a billiard ball is different than the point-wise curvature of the earth, since they are the same shape, just different sizes. It is the global Gauss-Bonnet theorem that validates this intuition. Although the curvature of the earth is much smaller than the curvature of a billiard ball, the earth has much more area than the billiard ball does. The total curvature is the same for both. Adding a lot of a small quantity is the same as adding a little of a large one.

If we combine the global Gauss-Bonnet theorem with the classification theorem for orientable surfaces (chapter 17), we can draw some interesting conclusions. For instance, the sphere is the only closed surface with positive Euler number. So, if we have a surface with positive total curvature, then it must be homeomorphic to a sphere. Likewise, if the total curvature of a closed surface is zero, then it must be homeomorphic to a torus. Every other closed orientable surface (genus g surfaces where $g > 1$) must have negative total curvature.

Although neither Gauss nor Bonnet noticed this global version of the theorem, Wilhelm Blaschke (1885–1962) chose to name it in their honor in a textbook he wrote in 1921.[7] It was in his book that the proof of the global theorem appeared using the local theorem. The first proof of the global Gauss-Bonnet theorem dates back to 1888, when Dyck proved the theorem using completely different techniques.[8] Yet again we see the unexpected ways names get attached to theorems.

[‡]The global Gauss-Bonnet theorem states that the total curvature of a surface S is $\int_S k\,dA = 2\pi\chi(S)$.

In these last two chapters we saw the beautiful and unexpected relationship between topology and geometry. Not only is the Euler number a topological invariant, but it is the link that ties these two different fields together. We see yet another reason why Euler's formula is a fundamental relationship in mathematics. In the next two chapters we see that we can generalize the Euler number to higher-dimensional objects.

CHAPTER 22

NAVIGATING IN n DIMENSIONS

LISA: *Where is my dad?*
PROFESSOR FRINK: *Well, it should be obvious to even the most dimwitted individual who holds an advanced degree in hyperbolic topology that Homer Simpson has stumbled into the third dimension . . . [drawing on a blackboard] Here is an ordinary square—*
CHIEF WIGGUM: *Whoa, whoa. Slow down, egghead!*
PROFESSOR FRINK: *—but suppose we extend the square beyond the two dimensions of our universe, along the hypothetical z-axis, there [everyone gasps]. This forms a 3-dimensional object known as a "cube," or "Frinkahedron" in honor of its discoverer.*
—The Simpsons, "Treehouse of Horror, VI"

Thus far, all of our topological objects have been curves or surfaces—objects that are locally 1- or 2-dimensional, and which live in 2-, 3-, or 4-dimensional space. Surfaces were the topological generalization of polyhedra, and Euler's polyhedron formula generalized nicely to the Euler number for surfaces. At this point, it is natural to ask what we can say about higher-dimensional topological shapes. What are they, and is there an Euler number for them too?

As we will see in chapter 23, Poincaré defined the Euler number for higher-dimensional topological spaces and proved that it is a topological invariant. But before we discuss Poincaré's contributions, we should discuss dimension and some of the early attempts to extend the Euler number.

Everyone is familiar with 0-, 1-, 2-, and 3-dimensional spaces. Three-dimensional space is the environment in which we live. Trees, houses, people, and dogs are all 3-dimensional entities. 3-dimensional space contains 2-dimensional objects. A chalkboard, a piece of paper, and a television screen are all 2-dimensional. String, a balance beam, and a coiled telephone

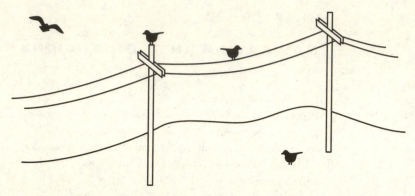

Figure 22.1. Birds experiencing dimensions 0, 1, 2, and 3.

cord are 1-dimensional. The period at the end of this sentence is zero dimensional.

It is common to associate dimension with geometric figures such as points, lines, and planes. However, as we saw in the previous chapters, we want a looser definition of dimension than these rigid geometric ones. A more healthy way to think about dimension is in terms of degrees of freedom—the dimension is the number of independent directions that an object can move.

Consider the flock of birds shown in figure 22.1. Each bird has limitations upon where it can move—they each have a different number of degrees of freedom. The bird on the top of the telephone pole cannot move anywhere. It is experiencing zero-dimensional space. The bird on the wire can move side-to-side. It has one degree of freedom, so it inhabits a 1-dimensional space. The bird on the ground is living in 2-dimensional space, and the flying bird is in 3-dimensional space. Notice that we did not speak of straight lines and flat planes, just degrees of freedom. The hanging wire is certainly not straight and the ground has hills and valleys.

Because we live in a 3-dimensional world, it is easy for us to conceptualize dimensions 0, 1, 2, and 3. Four or more dimensions lie outside our realm of experience. When we investigated cross caps, the Klein bottle, and the projective plane, we recognized the need for a fourth dimension. Although it is not easy to imagine this hop into the fourth dimension, it is not too difficult to visualize these surfaces. After all, these 4-dimensional surfaces are mostly 3-dimensional. Topological entities that require more than a brief detour into the fourth dimension are another story.

It is common to hear people say that time is the fourth dimension. This perspective is due to Joseph-Louis Lagrange and has been around since 1788.[1] Time is a quantity with which we are familiar, and it may help get a handle on 4-dimensional space. But there is a downside. We cannot ignore the "arrow of time." The three physical dimensions that we encounter have no direction. Particles can move back and forth on a line without contradicting the laws of physics. However, that same particle cannot go backward and forward in time. There is something inherently different about time compared to the other three dimensions. In general we do not want our fourth dimension to have this restrictive property.

In practice, high-dimensional spaces arise naturally. Predicting the motion of the space shuttle requires six dimensions—three dimensions for its position and three dimensions for its velocity. To pinpoint the positions and velocities of the sun, earth, and moon, we need eighteen. An economist's financial model, an ecologist's population study, and a physicist's quantum theory may each possess a large number of variables (each of which contributes a dimension). From a mathematical point of view we can add as many dimensions as we need.

Regardless of the source of our high-dimensional space, in our discussion we assume that each dimension is physical, that each is no different from the three usual dimensions. We are not asserting that there *are* more than three physical dimensions. Perhaps there are, and perhaps there are not (physicists studying string theory claim that there are at least ten dimensions). Mathematically speaking, this is irrelevant.

We denote n-dimensional Euclidean space by $\mathbb{R}^n$. $\mathbb{R}^1$ is the set of real numbers—the number line that we study in grade school. Each point can be represented by a single x-value. $\mathbb{R}^2$ is the infinite plane. It has an x- and a y-axis, and using these coordinate axes, each point can be represented by an ordered pair (x, y). Three-dimensional Euclidean space is denoted $\mathbb{R}^3$, and each point in $\mathbb{R}^3$ is uniquely represented by an ordered triple (x, y, z). Mathematically, it is trivial to extend these notions to n-dimensional Euclidean space. Each point in $\mathbb{R}^n$ can be uniquely described by an ordered n-tuple, $(x_1, x_2, \ldots, x_n)$. We can work with these high-dimensional spaces whether they exist physically or not.

We spent a lot of time learning about surfaces. We described surfaces as being locally two-dimensional. An ant living on the surface has two degrees of freedom. We can extend this same idea to higher-dimensional shapes. An *n-dimensional manifold*, or *n-manifold*, is a topological object that looks locally like n-dimensional Euclidean space. An inhabitant of one of these manifolds has n degrees of freedom. Just as was the case with surfaces,

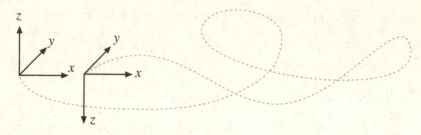

Figure 22.2. Coordinate axes in a nonorientable 3-manifold.

manifolds exhibit local simplicity and global complexity. They may have holes or other nontrivial topology. Regardless of the global characteristics, up close, all n-manifolds look like $\mathbb{R}^n$.

Also like surfaces, n-manifolds can be orientable or nonorientable. The easiest way to test for nonorientability in a manifold is by using Dyck's criteria (chapter 16). Suppose we had two identical coordinate frames in a nonorientable n-manifold. Then it is possible to move one coordinate frame around the manifold in such a way that when it returns to the first one, it is impossible to get all of the axes to line up correctly. For example, in a nonorientable 3-manifold, when the x- and y-axes are in alignment, the z-axes point in opposite directions (see figure 22.2).

Manifolds of any dimension can have boundaries, and the boundary of an n-manifold is a manifold of one dimension less. The boundary of a 1-manifold is 0-manifold (two points), the boundary of a 2-manifold (a surface) is a 1-manifold (one or more circles), and the boundary of a 3-manifold (a solid) is a surface. For example, the boundary of a solid torus is the usual (hollow) torus. The boundary of a solid ball is a sphere, and more generally, the n-ball, B^n, is an n-manifold and its boundary is the $(n-1)$-sphere, S^{n-1} (see chapter 19 for the definitions of S^n and B^n).

The history of manifolds dates back to Riemann and his study of multivalued complex functions and the associated Riemann surfaces. But it was near the turn of the twentieth century that Poincaré argued that the manifold was an important object of study, and he gave several ways of describing it. Probably the simplest way is by expressing it as a subset of $\mathbb{R}^n$ using an equation or equations. For example, $x^2 + y^2 + z^2 = 1$ is a sphere and $(3 - \sqrt{x^2 + y^2})^2 + z^2 = 1$ is a torus. They both reside in $\mathbb{R}^3$.

Sometimes Poincaré presented a manifold as an n-dimensional polyhedron called a *simplicial complex*. In a simplicial complex, the generalization of a vertex, an edge, and a face is a *simplex*. We may assume that all of the

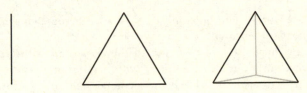

Figure 22.3. 0-, 1-, 2-, and 3-simplices.

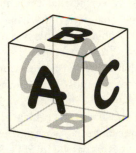

Figure 22.4. Glue the corresponding faces to obtain a 3-torus.

simplices are triangles, or the higher-dimensional analogue of a triangle. As we see in figure 22.3, a *k-simplex* is the k-dimensional figure determined by $k+1$ points. A 0-simplex is a point, a 1-simplex is a line segment, a 2-simplex is a triangle, a 3-simplex is a triangular pyramid, and so on. In a simplicial complex we assume that when two neighboring simplices meet, they meet along some lower-dimensional simplex. (Note that just as Hessel's polyhedra [chapter 15] were not surfaces, not every simplicial complex is a manifold.)

Yet another way Poincaré described manifolds was by generalizing Klein's construction of surfaces. Just as Klein created surfaces by gluing together sides of polygons, so did Poincaré create n-manifolds by gluing together the faces of n-dimensional polyhedra. We obtain a torus by gluing opposite sides of a square together without any twists. Likewise, we construct a 3-torus (the 3-dimensional analogue of a torus) by gluing opposite sides of a cube together in pairs with no twists (see figure 22.4). A 3-torus is an example of a closed, orientable 3-manifold.

The abstract definition of manifold does not specify where the manifold "lives." We were able to define and understand the properties of the Klein bottle without ever knowing that it cannot exist in $\mathbb{R}^3$. We ask, given a generic n-manifold, can we always place it in some Euclidean space, $\mathbb{R}^m$, so that it has no self-intersections? If so, how big does m have to be? Hassler Whitney proved that any n-manifold can be placed in some Euclidean space with dimension at most $2n$. This has become known as the *Whitney embedding theorem*.

In chapter 17 we encountered the classification theorem for surfaces. Every surface is either a sphere with handles or a sphere with cross caps. It is reasonable to ask whether it is possible to classify n-manifolds for n greater than two. It turns out that this is an extremely challenging problem. In chapter 17 we asserted that the dimension of an n-manifold is a topological invariant—that it is impossible for a 5-manifold and a 7-manifold to be homeomorphic. Even this was a difficult result to justify. It was not until 1911 that Brouwer proved the *invariance of dimension* theorem,[2] which states that $\mathbb{R}^n$ is not homeomorphic to $\mathbb{R}^m$ when $m \neq n$. Later we will discuss one of the most famous classification questions, one whose correct solution is worth $1 million.

The importance of the classification problem should not be understated. One of the basic open questions is, what is the shape of the universe? String theorists aside, it appears that we live in a 3-dimensional universe— a gigantic 3-manifold (presumably without boundary!). What are the properties of this manifold? Is its diameter finite or does it extend forever? Is it topologically the same as $\mathbb{R}^3$ or does it have some nontrivial topological features. Even more bizarre, could it be nonorientable? Is it possible for a right-handed astronaut to fly away from earth, and return left-handed?

Now that we have the notion of manifolds of all dimensions, it is natural to ask whether we can apply Euler's formula to them. To do so, we should return to polyhedra. Cauchy was the first to catch a glimpse of the higher-dimensional generalization of Euler's formula.[3] In the same paper in which he proved Euler's formula by transporting a polyhedron onto a plane, he stated and proved a higher-dimensional analogue of Euler's formula. He argued that if vertices, edges, and faces are inserted into the interior of a convex polyhedron, dividing it into S convex polyhedra, and if the total number of vertices, edges, and faces (including those in the interior) is V, E, and F, respectively, then

$$V - E + F - S = 1.$$

To illustrate Cauchy's theorem, look at the decompositions of the octahedron and the cube in figure 22.5. A new face in the interior of the octahedron divides it into two polyhedra, so $S = 2$. There are 6 vertices, 12 edges, and 9 faces. As Cauchy asserted, $6 - 12 + 9 - 2 = 1$. Likewise, the cube is divided into 3 polyhedra with 12 vertices, 22 edges, and 14 faces, and $12 - 22 + 14 - 3 = 1$.

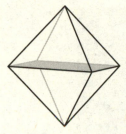
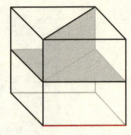

Figure 22.5. Decompositions of an octahedron and a cube.

In 1852 Ludwig Schläfli discovered a version of Euler's formula that held for convex polyhedra of all dimensions, but his work was not published in full until 1901 and by then his results had been rediscovered by others.[4] Suppose that P is an n-dimensional polyhedron that has b_0 vertices, b_1 edges, b_2 faces, and in general b_k facets of dimension k. Schläfli thought of these n-dimensional polyhedra as hollow shells bounded by $(n-1)$-dimensional faces; in this case that means $b_n = 0$. Define the Euler number by extending the alternating sum to facets of higher dimension: $\chi(P) = b_0 - b_1 + b_2 - \cdots \pm b_{n-1}$. Schläfli observed that $\chi(P) = 0$ when n is odd and $\chi(P) = 2$ when n is even.

Let us examine Cauchy and Schläfli's results from a modern topological viewpoint. First of all, both Cauchy and Schläfli were considering convex polyhedra, which do not have holes or tunnels. Topologically, Schläfli's hollow n-dimensional polyhedron is homeomorphic to the $(n-1)$-dimensional sphere, S^{n-1}. Thus, Schläfli's theorem shows that $\chi(S^n) = 0$ when n is odd and $\chi(S^n) = 2$ when n is even. On the other hand, Cauchy assumed his convex polyhedron was solid, so it is topologically the same as the 3-ball, B^3. Cauchy proved that $\chi(B^3) = 1$, and we now know that $\chi(B^n) = 1$ for all n. To see this, create B^n by "filling in" Schläfli's polyhedron with one n-dimensional facet. So $\chi(B^n) = \chi(S^{n-1}) + (-1)^n$. For n even, $\chi(B^n) = 0 + 1 = 1$; and for n odd, $\chi(B^n) = 2 - 1 = 1$.

The next higher-dimensional generalization was given by Listing. We have met Listing several times already. He contributed to graph theory (chapter 11), he was the first mathematician to study knots (chapter 18), and he discovered the Möbius band before Möbius and even coined the term topology (chapter 16). In fact, he was the first to have a truly topological view of Euler's formula and was one of the first mathematicians to *think* like a topologist. One might expect him to be regarded as one of the giants of topology. In reality, he was largely ignored in his lifetime and remained

Figure 22.6. Johann Listing.

an obscure figure for years after he died. Even now, the *Dictionary of Scientific Biography*, an eighteen-volume set containing short biographies of the most important scientists and mathematicians in history, contains no entry for Listing.

It is not clear why he never gained his rightful place in history. His academic pedigree was excellent. He was a doctoral student of Gauss, and he remained in his mentor's inner circle until the day Gauss died (Listing was actually present at his death). For eight years he lived next door to Riemann. (Surprisingly, there is no evidence of collaboration or meaningful conversation between these two who could have shared so much. It has been suggested that Listing may have feared that the tuberculosis that had ravaged Riemann's family was contagious.[5]) Listing made important contributions in areas of science as well, such as the optics of the eye. He coined several terms other than topology that have persisted to this day, such as "micron," the unit for one millionth of a meter.

Perhaps his obscurity was due to his personal attributes. Although he was a gregarious and kind individual, he suffered from manic depression, he was perpetually in financial danger due to his extensive debts, and his wife was often in trouble with the law. Maybe it was his restless spirit that took him away from mathematics for years at a time, his bad career decisions, or his refusal to play the political games in academia. It might have been his way of presenting mathematics. His writing was always too

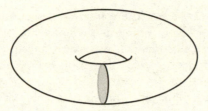

Figure 22.7. A decomposition of a solid torus.

focused on details, making it difficult to perceive the important and deep mathematics that he discovered.

He wrote two monographs on topology, one in 1847 and one in 1861.[6] The first, the previously mentioned *Topologie*, consisted mostly of his topological musings. The second, with the long title of *Der Census räumlicher Complexe oder Verallgemeinerung des Euler'schen Satzes von den Polyedern* (*The Census of Spatial Complexes or Generalization of Euler's Theorem on Polyhedra*), contained his generalizations of Euler's formula to nonconvex 3-dimensional shapes. In 1884 P. G. Tait lamented that Listing's writings on topology

> have not yet been rescued from their most undeserved obscurity, and published in an English dress, especially when so much that is comparatively worthless, or at least not so worthy, has already secured these honours.[7]

In *Census* Listing abandoned the rigid polyhedral view of these shapes; instead he gave the problem a topological treatment. Listing counted the number of vertices, edges, faces, and (3-dimensonal) spaces, but he allowed these features to have nontrivial topology, or *cyclosis* (as he called it). For instance, he counted a circle as an edge and a sphere as a face, but when he counted them he modified the tally by taking into account their topology. He counted a cylinder as a face, but because it contained a nontrivial loop, he subtracted one. Thus, if A, B, C, and D, are the numbers of vertices, edges, faces, and spaces, each suitably purged of its cyclosis, then $A - B + C - D = 0$.

To give a flavor of how Listing's decomposition works, we apply it to the solid torus with the decomposition shown in figure 22.7. It has no vertices, it has a single circular edge, it has two faces (one shaped like a cylinder and one a disk), and it has two spaces (the inside of the cylinder and the ambient space, which he always counted). Because this decomposition has no vertices, $A = 0$. There is one edge, but it contains a closed loop, so $B = 1 - 1 = 0$. There are two faces, but since the cylindrical one contains a closed loop around its circumference, C decreases by one. So $C = 2 - 1 = 1$. Finally, there are two spaces, but since the exterior space contains a nontrivial loop, we have $D = 2 - 1 = 1$. As predicted by Listing, $A - B + C - D = 0 - 0 + 1 - 1 = 0$.

Listing's approach to the problem was remarkably clever and insightful. It was the first attempt to treat the 3-dimensional Euler's formula in a truly topological way. However, it was far from perfect. At the very least, this means of computing A, B, C, and D was confusing. Listing abandoned all of the beautiful simplicity of Euler's vertices, edges, and faces. Instead, we must understand the topology of each building block of Listing's partition.

The next major addition to the theory of n-dimensional topology was made by Riemann and the Italian mathematician Enrico Betti (1823–1892). In order to understand their contributions, we must return to Riemann's study of surfaces.

In his 1851 doctoral dissertation Riemann presented a topological invariant that counted the number of holes in an orientable surface. He called it the *connectivity number* of the surface.[8] A surface (with or without boundary) has connectivity number n, or is *n-connected*,* if n is the largest number of cuts that one can make without disconnecting the surface.† If the surface has a boundary, then the cuts must begin and end at the boundary. If the surface has no boundary, then the first cut must begin and end at the same point (thereafter the surface has a boundary).

In figure 22.8 we have three surfaces with boundary: a cylinder, a disk with three holes, and a Möbius band. The dashed lines represent the cuts. The cylinder and the disk with holes have connectivity numbers 1 and 3, respectively. Riemann's work on connectivity numbers predates Möbius's discovery of nonorientable surfaces, but we can compute the connectivity numbers for nonorientable surfaces in exactly the same way. So the Möbius band is 1-connected.

The simplest closed surface is the sphere. Any cut along a closed curve will disconnect the sphere. So it is 0-connected. If we cut a torus once around the tube, then it becomes a cylinder. We may then cut it a second time along the length of the cylinder, obtaining a rectangle. The

---

*Today the term *n-connected* means something slightly different.

†Actually, Riemann's connectivity number was one larger than this value, but we will keep the smaller value to be consistent with modern notation.

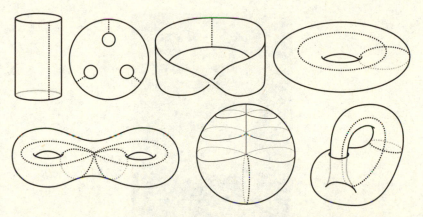

Figure 22.8. Cuts to determine the connectivity numbers for various surfaces.

connectivity number of the torus is 2. Similarly, we can compute the connectivity numbers of other surfaces. The double torus is 4-connected, the projective plane is 1-connected, and the Klein bottle is 2-connected.

Although it may seem that the connectivity number is an important new topological invariant, it is actually the Euler number in disguise. The astute reader may have noticed the relationship between the connectivity number and the genus for orientable closed surfaces. Indeed, Riemann noticed this as well—the connectivity number is simply twice the genus. If we know the genus, the Euler number, or the connectivity number, then we know the other two.

Let us be more precise about the relationship between the connectivity number and the Euler number. Imagine that before we do any cutting, we draw the cut-lines on an n-connected surface, S. This gives a very simple partition of the surface. For the sake of simplicity, we may assume that each cut begins and ends at the same point, so $V = 1$. At the end of our cutting we have a single face, so $F = 1$. Moreover, each cut-line is an edge, so $E = n$. This gives the following simple relationship between the connectivity number and the Euler number:

$$\chi(S) = 1 - n + 1 = 2 - n.$$

Toward the end of his life, Riemann's health began to decline. Between 1862 and his death in 1866, he made several trips to Italy for convalescence. While he was there, he visited his friend Betti, whom he had met in Göttingen in 1858. Betti was a university professor at the University of Pisa. He had also taught high school, was a member of Parliament, and had

Figure 22.9. Enrico Betti.

served as a senator. He was a renowned mathematician and gifted teacher who was partially responsible for the rebirth of mathematics in Italy after the reunification.

During his visits to Italy, Riemann spoke with Betti about how to extend the connectivity numbers to manifolds of higher dimension. It is difficult to say who contributed what to the theory. It was Betti who, in 1871, published these generalizations, but letters and notes show that Riemann knew much of this as early as 1852.

The idea behind the generalizations is that, just as Riemann counted the maximal number of 1-dimensional cuts for a surface, in an n-dimensional manifold we may also count the maximal number of m-dimensional manifolds (subject to somewhat complicated criteria) for each $m \leq n$. This gives a connectivity number b_m for each m from 0 to n. In this notation, b_1 is Riemann's connectivity number.

Betti proved that these b_m were topological invariants for the manifold. However, working with n-manifolds is tricky, and later it became apparent that there were subtle errors with Betti's arguments and definitions. Nonetheless, Betti's work was an extremely important step toward understanding the topology of n-manifolds.

It was Henri Poincaré who aimed to fix Betti's errors. He did that, and much more.

CHAPTER 23

HENRI POINCARÉ AND
THE ASCENDANCE OF TOPOLOGY

*The mathematician's patterns, like the painter's or the poet's,
must be beautiful; the ideas, like the colors or the words, must
fit together in a harmonious way. Beauty is the first test: there
is no permanent place in this world for ugly mathematics.*
—G. H. Hardy[1]

If Euler's theorems on the bridges of Königsberg and polyhedra mark
the birth of topology, and the contributions of Listing, Möbius, Riemann,
Klein, and other nineteenth-century mathematicians signify topology's
adolescent years, then its coming of age was signaled by the work of
Henri Poincaré. Before this there were theorems that we now categorize as
topological, but it was not until the waning years of the nineteenth century
that Poincaré systematized the field.

Looking back at his complete body of work, we see a common theme:
a topological view of mathematics. Perhaps this qualitative approach to
the subject came from his distaste for (or as he said, his difficulty with)
performing mathematical calculations. Perhaps it was in response to his
notorious lack of artistic talent (remember, he called geometry "the art of
reasoning well on badly made figures"). Regardless of the cause, Poincaré
eventually recognized this common trait, and wrote, "Every problem I had
attacked led me to *Analysis Situs*."[2]

Poincaré was referring to the seminal 123-page article *Analysis Situs*[3]
that he wrote in 1895. In the next decade he followed up with five
groundbreaking sequels, or complements as he called them.[4] About these
six papers, Jean Dieudonné wrote:

> As in so many of his papers he gave free rein to his imaginative
> powers and his extraordinary "intuition," which only very seldom
> led him astray; in almost every section is an original idea. But we

should not look for precise definitions, and it is often necessary to guess what he had in mind by interpreting the context. For many results, he simply gave no proof at all, and when he endeavored to write down a proof hardly a single argument does not raise doubts. The paper is really a *blueprint* for future developments of entirely new ideas, each of which demanded the creation of a new technique to put it on a sound basis.[5]

One imagines a Johnny Appleseed, wandering through a barren land, scattering seeds that would later grow into fruit-bearing trees. It is hardly an exaggeration to say that nearly all of the research in topology until the early 1930s came from the work of Poincaré.

One of his contemporaries wrote, "In the domain of *Analysis Situs* Poincaré has recently brought us an abundance of new results, but at the same time raised an abundance of new questions which still await settlement."[6] The gaps and holes in Poincaré's arguments were real, and they took time to sort out. The intuitive approach to the subject by Poincaré and his predecessors needed to be supported by solid mathematical arguments. Rigor and a uniform standard for proof in topology appeared around 1910, and it took several decades to build a solid structure from Poincaré's blueprints.

One of Poincaré's many important contributions was his invention of *homology*. This was an ingenious way of formalizing the study of Riemann's connectivity numbers and Betti's higher-dimensional generalizations. Today homology is the primary means of analyzing manifolds. He introduced homology in *Analysis Situs* and refined it in each of the complements. It took approximately thirty years for the theory of homology to take its modern form.

It is beyond the scope of this book to describe homology theory, either in modern terms or in Poincaré's. Instead, we will settle for a superficial account with a focus on intuition. Rather than give the n-dimensional version, we will discuss 1-dimensional homology on surfaces.

One way to view 1-dimensional homology is to look at loops drawn on a surface. We will not focus on a fixed loop, but we will let it move around on the surface. It can stretch, shorten, and wiggle as needed, just as long as it does not break or leave the surface.

The simplest possible loop is one that is topologically trivial—one that can be shrunk back to a point. It may zig and zag around the surface wildly, but it does not go around any holes. For instance, because a sphere has no

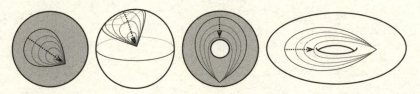

Figure 23.1. A disk and a sphere are simply connected, but an annulus and a torus are not.

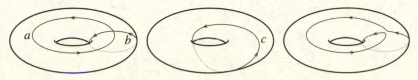

Figure 23.2. The cycle c is can be deformed into the cycle $a + b$.

holes, any loop drawn on the surface of the sphere can be shrunk to a point. The simplest surfaces are those for which, like the sphere, every loop is topologically trivial. Such a surface is called *simply connected*. As we see in figure 23.1, a disk and a sphere are simply connected, but an annulus and a torus are not.

We know from the classification of surfaces that the sphere is the only simply connected closed surface. All others have infinitely many nontrivial loops. Poincaré realized that it was important to count the essential, or independent, nontrivial loops on the surface. For orientable surfaces, he called this number the 1-dimensional *Betti number*, in honor of Betti. In order to compute this number, he defined a strange arithmetic on loops, which we write as addition.

In homology, every loop has an orientation and is called a *cycle*. So the cycles a and $-a$ are the same loop, but with opposite orientations. The sum of two loops a and b is the union of the two cycles, thus $a + b$ and $b + a$ are the same, or with the notation we will use, $a + b \equiv b + a$. At times we may want to think of $a + b$ as a loop itself. According to our arithmetic we could follow loop a then b or loop b then a. Although these may be different loops, they represent the same cycle. We let two cycles with opposite orientations cancel, so $a + (-a) + b \equiv b$. Furthermore, if the cycle a can be deformed into the cycle b, then $a \equiv b$.

To get a feeling for how this addition works, consider the three cycles a, b, and c on the torus in figure 23.2. As we can see, it is possible to deform the cycle c so that it coincides with the cycle a followed by the cycle b. Thus c and $a + b$ are the same cycle, or $c \equiv a + b$.

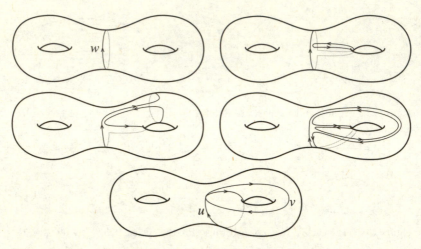

Figure 23.3. A zero cycle on the double torus that cannot be shrunk to a point.

If this is truly like addition, there must be a zero cycle. What does a zero cycle look like? The most obvious zero cycle is one that can be shrunk to a point. On a simply connected surface every cycle is zero. Is that it? Are the topologically trivial loops the only zero cycles? They are not. The loop w in figure 23.3 runs around the waist of the double torus and cannot be shrunken to a point. However, we can deform it so that it traces out the cycle u, then v, then $-u$, then $-v$. So, $w \equiv u + v + (-u) + (-v) \equiv 0$.

We see that the cycle c in figure 23.2 can be written as a sum of cycles a and b. It turns out that any cycle on the torus can be written as a sum of a and b. In other words, given any cycle d on the torus, it is possible to find integers m and n so that $d \equiv ma + nb$. In essence, a and b are the only cycles that matter, so, according to Poincaré, the 1-dimensional Betti number of the torus is 2. Likewise, for the double torus in figure 23.3 the loops u and v are essential, and there are two more around the other handle. The 1-dimensional Betti number of the double torus is 4.

For orientable surfaces the number of such loops is the first Betti number, but for nonorientable surfaces, something strange happens. In our experience, if we saw the equation $a + a = 0$, then we would conclude that $a = 0$. This is always the case for real numbers. However, for cycles it may be the case that $a \not\equiv 0$, but $a + a \equiv 0$. In fact, this is not so foreign to us. Many cars have odometers that go up to 99,999 miles. According to such an odometer, $50,000 + 50,000 = 0$. Another example is a clock running

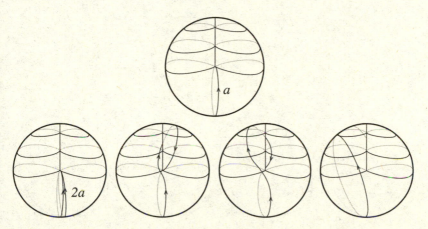

Figure 23.4. For the projective plane, $2a \equiv 0$.

on military time. Midnight is 0:00, noon is 12:00, and immediately before midnight is 23:59. So, 12 hours after 12:00 is 0:00, or $12 + 12 = 0$.

To see this strange arithmetic in action we return to the projective plane and the Klein bottle. We saw in figure 22.8 that the connectivity number of the projective plane is 1. Let us denote the corresponding cycle by a and give it an orientation as in figure 23.4. Now $a + a$, or $2a$ for short, is the cycle tracing out a two times. Remarkably, as we can see in the figure, this doubled cycle is indeed topologically trivial—by manipulating the loop it is possible to shrink it to a point. So $2a \equiv 0$.

The same thing occurs on the Klein bottle, but the rationale is slightly different. Earlier we saw that the connectivity number of the Klein bottle is 2. Denote the corresponding cycles (with orientation) b and c, as in figure 23.5. As we can see, the doubled cycle $2b$ is equivalent to $b + c + (-b) + (-c)$. In other words, even though $2b$ is not topologically trivial, $2b \equiv 0$.

So we may separate these essential cycles into two classes, depending on whether or not they exhibit this behavior. We continue to call the number of cycles that do not have this behavior the 1-dimensional Betti number. If the surface has a cycle a for which $na \equiv 0$ (and n is the smallest positive integer for which this is true), then we call n a *torsion coefficient* for the surface. Thus, in dimension 1 the projective plane has Betti number 0 and a torsion coefficient of 2, while the Klein bottle has Betti number 1 and torsion coefficient 2.

In a similar manner Poincaré defined the higher-dimensional Betti numbers and torsion coefficients, except that instead of having loops as

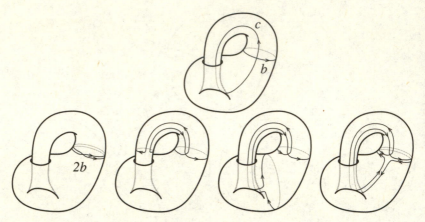

Figure 23.5. For the Klein bottle, $2b \equiv b + c + (-b) + (-c) \equiv 0$.

TABLE 23.1:
Betti numbers and torsion coefficients of surfaces.

Surface S	$\chi(S)$	b_0	b_1	b_2	Torsion coefficients (in dimension 1)
Sphere	2	1	0	1	none
Torus	0	1	2	1	none
2-holed torus	−2	1	4	1	none
g-holed torus	$2 - 2g$	1	$2g$	1	none
Klein bottle	0	1	1	0	2
Projective plane	1	1	0	0	2
Sphere with c cross caps	$2 - c$	1	$c - 1$	0	2

cycles he used higher-dimensional manifolds. Poincaré proved that both the Betti numbers and the torsion coefficients are topological invariants for manifolds. In table 23.1 we show the Betti numbers and torsion coefficients for closed surfaces. We let b_i denote the ith Betti number.[*]

In *Analysis Situs* Poincaré followed the ideas of Riemann and Betti. In response to a call for rigor, he changed directions in his later papers. It was at this point that he began working with simplicial complexes, the n-dimensional generalization of polyhedra. In this context, the cycles in homology were built from the features on the polyhedron. For example,

---

[*] Poincaré followed Riemann's convention and took the ith Betti number to be one larger than this value, but for simplicity we will use the modern convention.

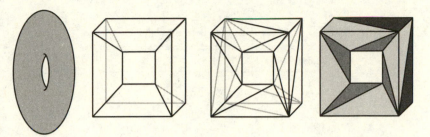

Figure 23.6. A simplicial complex for a solid torus.

a 1-dimensional cycle was not an arbitrary loop on the manifold, but sequences of edges on the polyhedron that formed a loop.

On a practical level, simplicial complexes are much easier to work with than Poincaré's first model. Poincaré could describe the complex in terms of an *incidence matrix*, a rectangular array of numbers that showed which simplices were neighbors. Using these matrices, it was purely a mechanical process to compute Betti numbers and torsion coefficients.

With this higher-dimensional generalization of a polyhedron, it is natural to ask whether the Euler number can be extended to higher-dimensional manifolds. Indeed, as Cauchy and Schläfli did before him, Poincaré generalized the Euler number by computing the alternating sum of the number of k-simplices. In other words, if a manifold M can be partitioned as a simplicial complex with a_k simplices of dimension k, then he defined the Euler number to be

$$\chi(M) = a_0 - a_1 + a_2 - \cdots \pm a_n.$$

This generalization of the Euler number (or Euler characteristic) to n-dimensional space is called the *Euler-Poincaré characteristic* of M.

For example, a solid torus is a 3-manifold with boundary (the boundary is the torus, a 2-manifold). In figure 23.6 we show how to partition the solid torus as a simplicial complex. It has 12 vertices (0-simplices), 36 edges (1-simplices), 36 faces (2-simplices), and 12 triangular pyramids (3-simplices). So, $a_0 = 12$, $a_1 = 36$, $a_2 = 36$, and $a_3 = 12$, and the Euler-Poincaré characteristic is $\chi(\text{solid torus}) = 12 - 36 + 36 - 12 = 0$.

Just as the Euler number is a topological invariant for surfaces, so is the Euler-Poincaré characteristic an invariant for n-manifolds. To prove this, Poincaré showed something much more interesting. He proved that if the

TABLE 23.2:
The symmetry of Betti numbers.

Manifold	Betti numbers: $b_0, b_1, b_2, \ldots, b_n$
S^1	1, 1
S^2	1, 0, 1
S^n	1, 0, \ldots, 0, 1
Torus	1, 2, 1
Double torus	1, 4, 1
g-holed torus	1, 2g, 1

kth Betti number is b_k, then

$$\chi(M) = b_0 - b_1 + b_2 - \cdots \pm b_n.$$

That is, to compute the Euler-Poincaré characteristic, we throw out the torsion coefficients and take the alternating sum of the Betti numbers! In table 23.1 we can see that this equality holds for the Betti numbers for surfaces. Because each of the b_k are topological invariants, so is the alternating sum. So the Euler-Poincaré characteristic is a topological invariant.

In 1895 Poincaré discovered a beautifully symmetric relation among the Betti numbers.[7] The sequence of Betti numbers for several manifolds are shown in table 23.2. He noticed that the Betti numbers came in pairs, with the first Betti numbers the same as the last ones: $b_0 = b_n$, $b_1 = b_{n-1}$, etc. This became the celebrated Poincaré duality theorem.

> POINCARÉ DUALITY THEOREM
> If $b_0, b_1, \ldots, b_n$ are the Betti numbers for a closed orientable n-manifold, then $b_i = b_{n-i}$ for all i.

We encountered duality earlier when we discussed Kepler's pairing of the Platonic solids (chapter 6). The use of the term "duality" in both cases is not a coincidence; Kepler's observation is Poincaré duality in disguise. Poincaré duality states that when computing Betti numbers for a manifold, we are free to exchange the roles of i-dimensional and $(n-i)$-dimensional simplices. The duality of the Platonic solids illustrates this behavior. For instance, the icosahedron gives a partition of the sphere into vertices, edges, and faces. When we use Kepler's notion of duality to turn every vertex of the icosahedron into a face and every face into a vertex, we get the dodecahedron, another partition of the sphere.

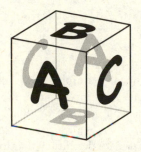 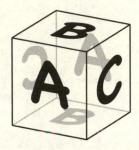

Figure 23.7. The 3-torus and a nonorientable 3-manifold.

In *Analysis Situs* Poincaré wrote, "This theorem has not, I believe, ever been stated; it is, however, known to many, who have even found some applications of it."[8] We do not know who these other people were nor how they used the relationship, but Poincaré's primary use was to prove the surprising fact that the Euler-Poincaré characteristic of any closed orientable manifold of odd-dimension is zero!

Indeed, consider the 3-torus, the 3-manifold obtained by gluing together the sides of the left-hand cube in figure 23.7. It has Betti numbers $b_0 = 1$, $b_1 = 3$, $b_2 = 3$, and $b_3 = 1$ (we will not prove this), so the Euler-Poincaré characteristic is,

$$\chi(\text{3-torus}) = 1 - 3 + 3 - 1 = 0.$$

In general, suppose M is any closed orientable manifold of odd dimension n. Poincaré duality and the alternating sum of the Euler-Poincaré characteristic give us pairings of equal Betti numbers with opposite signs. They cancel to give zero:

$$\begin{aligned}
\chi(M) &= b_0 - b_1 + b_2 - \ldots - b_{n-2} + b_{n-1} - b_n \\
&= (b_0 - b_n) - (b_1 - b_{n-1}) + \cdots \pm (b_{(n-1)/2} - b_{(n+1)/2}) \\
&= 0 - 0 + 0 - \cdots \pm 0 \\
&= 0.
\end{aligned}$$

It turns out that the Euler-Poincaré characteristic of any nonorientable closed manifold of odd dimension is also zero. We omit the complicated justification, but we will illustrate with an example. The 3-manifold obtained by gluing together the sides of the right-hand cube in figure 23.7 is nonorientable. It has Betti numbers $b_0 = 1$, $b_1 = 2$, $b_2 = 1$, and $b_3 = 0$ (it also has torsion in dimensions 1 and 2). We can see that Poincaré duality does not hold, yet the Euler-Poincaré characteristic is still zero:

$$\chi(M) = 1 - 2 + 1 - 0 = 0.$$

We should point out that odd-dimensional manifolds with boundary need not have zero Euler-Poincaré characteristic. For instance, the Euler-Poincaré characteristic of the n-dimensional ball, B^n, is 1 for all n.

It was during the first three decades of the twentieth century that topology made the transition from a discipline built on intuitive arguments to one based on rigorous proof. That is to say, it was during this time that topologists removed the gaps, holes, unwarranted assumptions, and errors from Poincaré's brilliant work.

For example, consider the following two assumptions made by Poincaré. First, he asserted that every manifold could be expressed as a simplicial complex, or more concisely, that every manifold can be triangulated. Second, he assumed that the *Hauptvermutung* held for arbitrary manifolds. (Recall that the *Hauptvermutung* says that any two partitions of a manifold can be refined by adding simplices so that they are topologically the same.) It turns out that in general, neither of these assumptions is true for general manifolds. Yet later mathematicians showed that Poincaré's conclusions still hold.

A significant improvement to Poincaré's ideas was made by the German mathematician Emmy Noether (1882–1935). Noether, the daughter of a mathematician, is remarkable for the prejudices she was forced to overcome. She was a woman in a highly male-dominated field. In 1904 the University of Erlangen permitted women to enroll, but until then she was only able to audit classes. She completed her PhD in 1907. In 1915, after her reputation as a first-class research mathematician began to spread, Klein and David Hilbert (1862–1943) brought her to Göttingen, intent on having her on the faculty. But it was not until 1919 that she was permitted to hold this position; in the meantime her courses were advertised under Hilbert's name, with her listed as an assistant. When the Nazis came to power, life changed for many Germans. In 1933 the Jewish Noether was forced to leave Göttingen for Bryn Mawr College in the United States. She died two years later.

She is most well-known for her pioneering work in the field of abstract algebra. Broadly speaking, abstract algebra is the study of sets with one or more binary operations (such as addition and multiplication and their inverses, subtraction and division).

Figure 23.8. Emmy Noether.

Until the middle of the 1920s homology was described in terms of Betti numbers and torsion coefficients. It took the algebraist Noether to recognize that there was significantly more structure to homology. She discerned that the key feature in homology was the ability to add and subtract cycles. Focusing on this arithmetic, she observed that homology was a specific instance of an algebraic entity called a *group*, and that *Betti groups*, or *homology groups* as they are now called, were the right way to view homology. In his autobiography, Pavel Aleksandrov wrote, "I remember a dinner at Brouwer's in [Noether's] honour during which she explained the definition of the Betti groups of complexes, which spread around quickly and completely transformed the whole of topology."[9]

Suddenly topologists had a brand-new tool kit available to them. All of the techniques and theorems from group theory were at their disposal. Powerful theorems could be proved without reinventing the wheel. Betti numbers and torsion coefficients fell out of the machinery in an obvious way, and the invariance of the Euler-Poincaré characteristic became a simple proof. In Aleksandrov's memorial address for Noether, he wrote:

> These days it would never occur to anyone to construct combinatorial
> topology in any way other than through the theory of ... groups;
> it is thus all the more fitting that it was Emmy Noether who first
> had the idea of such a construction. At the same time she noticed
> how simple and transparent the proof of the Euler-Poincaré formula
> becomes if one makes systematic use of the concept of Betti group.[10]

Yet again we see the power that results from the merging of different branches of mathematics. Descartes used analysis to understand geometry. Riemann and Poincaré used topology to understand analysis. Gauss and Bonnet used topology to understand geometry. Now topologists are free to use algebra to understand topology. This cross-fertilization has been extremely fruitful.

The incorporation of algebra into topology is so important that this entire area of topology—virtually all of the topology we have discussed in this book—is now called *algebraic topology*. In the decades after Poincaré's work, algebraic topology expanded beyond homology groups to include many other algebraic structures. Today, most topologists are algebraic topologists.

Epilogue: The Million-Dollar Question

Mathematics, rightly viewed, possesses not only truth, but supreme beauty—a beauty cold and austere, like that of sculpture, without appeal to any part of our weaker nature, without the gorgeous trappings of painting or music, yet sublimely pure, and capable of a stern perfection such as only the greatest art can show.
—Bertrand Russell, "The Study of Mathematics"[1]

In the twentieth century topology rose to become one of the pillars of mathematics, sitting side by side with algebra and analysis. Many mathematicians who do not consider themselves topologists use topology on a daily basis. It is inescapable. Today, most first-year graduate students in mathematics are required take a full-year course in topology.

One way to gauge the importance of an academic field is to see the trail of awards for accomplishments in the discipline. There is no Nobel Prize in mathematics. The mathematical equivalent is the Fields Medal. Fields Medals have been awarded once every four years since 1936 (except during World War II). At each ceremony the medals are given to at most four mathematicians under the age of forty who made an outstanding contribution to mathematics. Of the forty-eight recipients, roughly a third were cited for their work in topology, and even more made contributions in closely related areas.

One specific topological problem is itself responsible for three Fields Medals. It was one of the most famous unsolved problems of the twentieth century—one that is still so important, and so difficult, that its resolution is worth $1 million to the mathematician who first proves it. The name of this thorny problem is the *Poincaré conjecture*.

The classification theorem for surfaces is one of the most elegant theorems in all of mathematics. It states that every surface is uniquely determined by its orientability, Euler number, and number of boundary components. Obviously, it would be nice to have a similar theorem for manifolds of every dimension, but this is a tall order. It is clear that if such a classification exists, the same checklist will not suffice since the Euler-Poincaré characteristic of every closed odd-dimensional manifold is zero (see chapter 23).

Poincaré dreamed of classifying higher-dimensional manifolds, but even classifying 3-manifolds was beyond his reach. The Poincaré conjecture was only the first step in this classification process.

The simplest closed n-manifold is the n-sphere, S^n. Poincaré was searching for a simple test to determine if a given n-manifold is homeomorphic to S^n. In 1900 he thought he had such a test. He proved that any n-manifold that had the same homology as S^n must be homeomorphic to S^n.[2] The homology of the n-sphere is especially simple. It has Betti numbers of 1 in dimensions 0 and n, 0 in all other dimensions, and it has no torsion.

Four years later he realized that his proof was flawed.[3] Not only did he discover his own error, but he found a remarkable counterexample to his assertion. He constructed a pathological 3-manifold that had the same homology as S^3, but was not homeomorphic to S^3. He created the manifold by gluing together opposite faces of a solid dodecahedron, each with a 36° clockwise twist.

An interesting and surprising feature of *Poincaré's dodecahedral space* is that even though its first Betti number is zero, it is not simply connected. That is, every loop is zero in homology, but there exist loops in the manifold that cannot be shrunk to a point. In figure 23.3 we saw an example of a nontrivial loop on the double torus that is zero in homology, but in the dodecahedral space *every* loop that cannot be shrunk to a point is trivial in homology.

From this exotic example, Poincaré realized that homology alone was not sufficient to characterize S^n, not even S^3. So he abandoned the n-dimensional question, focusing instead on 3-manifolds. He suspected that if all loops in a 3-manifold are topologically trivial, then the manifold must be homeomorphic to S^3. This became the now-famous Poincaré conjecture.[4]

THE POINCARÉ CONJECTURE
Every simply connected closed 3-manifold is homeomorphic to the 3-sphere.

Actually, in Poincaré's paper this statement was not a conjecture, but a question about whether it was true. He did not state his opinion on the direction it would go. Proving this theorem would be a far cry from classifying all 3-manifolds, but it would be an important first step.

Everyone likes a good challenge, and the Poincaré conjecture is as challenging as they come. It became one of a short list of problems—the four color theorem, Fermat's last theorem, the Riemann hypothesis—that attained mythical status. Just like these others, the Poincaré conjecture consumed those who worked on it. Countless young mathematicians entered the chase. As one journalist wrote, "Mathematicians speak of Poincaré's conjecture like Ahab expounding on the White Whale."[5] In the years since 1904 there have been many who claimed to have a proof. Until recently, every argument has possessed a flaw—sometimes a subtle mistake embedded in the middle of hundreds of pages of deep mathematics.

Eventually the conjecture was generalized to n-manifolds—every n-manifold that is sufficiently like the n-sphere must be homeomorphic to S^n. This generalization may seem ridiculously ambitious. How could we prove it for $n = 100$ if we are unable to prove it for $n = 3$? If I can't bench press 175 pounds, what makes me think I can lift 500? Shockingly, the conjecture is easier for large n! It is often the case that low-dimensional topology is more challenging than high-dimensional topology. Roughly speaking, having more dimensions to play with gives more freedom to move things around without collisions.

The fiery young topologist Stephen Smale (b. 1930) from the University of California, Berkeley, gave the first proof of the generalized Poincaré conjecture. In 1960 he verified the conjecture for an important class of manifolds of dimensions $n \geq 5$—the so-called smooth manifolds.[6]

Smale is quite a colorful character. He was a vocal critic of the Vietnam War and a vehement free speech activist. His protests, which included criticism of American foreign policy while visiting Moscow, earned him a subpoena by the House Un-American Activities Committee. Later he found himself in hot water over a six-month trip to Brazil paid for by the National Science Foundation. The science advisor to President Johnson wrote, "This blithe spirit leads mathematicians to seriously propose that the common man who pays the taxes ought to feel that mathematical creation should be supported with public funds on the beaches of Rio de Janeiro."[7]

The impetus for this statement was Smale's now-famous quote: "My best-known work was done on the beaches of Rio de Janeiro."[8] During his stay in Brazil, Smale not only proved the high-dimensional Poincaré conjecture, but he also discovered the *Smale horseshoe*, the template for chaotic dynamical systems.

Within two years Smale's results for $n \geq 5$ were generalized to manifolds without the smoothness assumption.[9] It seemed that the remaining cases would follow in no time. Then progress stalled. The $n = 4$ case did not yield until 1982, when it was proved by thirty-year-old Michael Freedman from the University of California, San Diego.[10] Then progress stalled again. Each dimension was more difficult than the one before it. Dimension 3, that of the original conjecture, remained impenetrable. More false proofs came and went. The problem seemed untouchable.

In 1998, Smale published a list of the eighteen most important unsolved problems in mathematics[11] (David Hilbert had done the same thing a century before). The classical Poincaré conjecture was on this list.

That same year the Clay Mathematics Institute offered a $1 million bounty for what they viewed as the seven most challenging unsolved problems in mathematics. The Poincaré conjecture was on this elite list. To win the prize a mathematician must give a proof of the theorem, and it must survive the intense scrutiny of the mathematical community for two years after it appears in print.

In 1982 Bill Thurston announced a plan to completely classify the geometry of all 3-manifolds. He theorized that every 3-manifold could be carved up into regions, each of which possess one of eight geometric structures.[12] This became known as Thurston's geometrization conjecture. With these eight building blocks it would be possible to understand the geometry and topology of all 3-manifolds. In particular, it would imply that the only simply connected closed 3-manifold is the 3-sphere. It would prove the Poincaré conjecture.

That same year, Richard Hamilton, a mathematician at Cornell University, began a program that he believed would prove Thurston's geometrization conjecture.[13] He introduced a means of taking any 3-manifold and, as if blowing up a balloon, continuously deform it into what he hoped would be a form that clearly fit the Thurston model. He made substantial progress toward this goal. Most experts expected that his techniques should work, but Hamilton and others were unable to rule out or adequately deal with singularities—parts of the manifold that did not get nicer over time, but instead pinched off to something worse.

In 2002 an unassuming Russian mathematician named Grigori "Grisha" Perelman (b. 1966) from the Steklov Institute in St. Petersburg surprised the mathematical community by posting to the Internet the first of three short, but extremely dense papers. The papers, which totaled only sixty-eight pages, claimed to finish Hamilton's two-decade-old research program.[14] In them he showed that certain singularities would never occur and others could be carefully eliminated. Taken together, they proved the geometrization conjecture, and in its wake, the classical Poincaré conjecture.

The mathematical community was skeptical—they had heard such proclamations before and the papers were extremely short on details—but they were guardedly optimistic. Perelman was a respected mathematician and he was carrying out Hamilton's well-regarded plan.

Perelman's arguments left much unsaid. Even the foremost experts in geometry and topology had difficulty assessing the legitimacy of the proofs. Independently of each other, three teams of mathematicians pored through his arguments, filling in the missing details.[15] The average length of each analysis was over three hundred pages. They did not discover any major errors.

By the end of 2006 it was generally believed that Perelman's proof was correct. That year, the journal *Science* named Perelman's proof the "Breakthrough of the Year."[16] Like Smale and Freedman before him, the forty-year old Perelman was tapped to be a Fields Medals recipient for his contributions to the Poincaré conjecture (in fact, Thurston also received a Fields Medal for his work that indirectly led to the final proof). The countdown for the $1 million prize had begun (some wonder if Perelman and Hamilton will be offered the prize jointly).

It may be that one of the grand mathematical peaks has been summited, just like Fermat's last theorem a decade before. The flag has been planted. One might assume that this accomplishment sounds the death knell for an area of mathematics. This most certainly is not the case. From atop this summit mathematicians behold a stunning vista of previously unseen peaks, all waiting to be climbed. Like Fermat's last theorem, the result itself may not be as important as the huge body of mathematics that was created in an effort to prove it.

Great mathematics begets more great mathematics. Euler's solution to the bridges of Königsberg problem and his proof of the polyhedron formula set in motion a voyage of discovery through many areas of beautiful mathematics that led to the creation of topology. The Poincaré conjecture is but one waypoint on this exciting journey. Topology is still a live and vibrant field of study.

A bizarre and unfortunate postscript to this otherwise wonderful story was the effect Perelman's proof had on his life. It started well. In April of 2003 he embarked on a brief speaking tour. His lectures were attended by

Andrew Wiles, John Forbes Nash Jr. (the subject of the Hollywood biopic *A Beautiful Mind*), John Conway, and other well-known mathematicians. But after he returned to Russia, the intense scrutiny of the mathematical community and the posturing of other mathematicians who wanted a share of his credit began to take its toll on him.[17]

The always-solitary Perelman became more reclusive. He wanted his work to speak for itself and did not want to be a part of the verification process. Eventually, his disenchantment with the mathematical community overwhelmed him and he left his academic position, stopped responding to correspondences, and by all accounts abandoned mathematics entirely. In an unprecedented move that shocked the scientific community, he declined to accept his Fields Medal.

At the end of the summer of 2006, Perelman was unemployed and living with his mother off her meager pension in a small apartment in St. Petersburg. When asked if he would accept the Clay Mathematics Institute's prize money he responded, "I'm not going to decide whether to accept the prize until it is offered."[18]

To many it is shocking that Perelman would decline the Fields Medal and possibly the prize money. But to him, solving the problem was the ultimate reward, fame and money were irrelevant. As he said, "If the proof is correct then no other recognition is needed."[19] Every researcher understands Perelman's pure love for his subject and the profound satisfaction of a breakthrough discovery. It is not impossible to imagine that overwhelming personal attention would tarnish the accomplishment.

Surely it was this same unadulterated love of mathematics that drove Pythagoras, Kepler, Euler, Riemann, Gauss, Poincaré, and the rest to toil countless hours in pursuit of the perfect theorem and the perfect proof. We can only imagine Perelman's elation when he realized that he had proved the Poincaré conjecture, or Euler's joy when he saw that $V - E + F = 2$.

As Poincaré so eloquently wrote, "The scientist does not study nature because it is useful; he studies it because he delights in it, and he delights in it because it is beautiful. If nature were not beautiful, it would not be worth knowing, and if nature were not worth knowing, life would not be worth living."[20]

ACKNOWLEDGMENTS

It is my name on the front of this book, but there are many more people without whom this book would not exist. I would like to take this opportunity to thank them.

First and foremost, I must thank my editor, Vickie Kearn. Her calm presence and encouraging words meant a lot to me as I wrote this, my first book. She and the entire staff at Princeton University Press were wonderful to work with.

It would have been impossible for me to write this book without knowing exactly what Euler wrote. I am extremely grateful to Chris Francese for helping me translate Euler's paper from the original Latin. Also, Anne Maiale, Tony Mixell, Sandra Alfers, Wolfgang Müller, and Lucile Duperron were helpful to me in getting other translations precisely right.

I appreciate Rich Klein, Ed Sandifer, Paul Nahin, Klaus Peters, Karl Qualls, my parents Gail and Frank Richeson, and the anonymous readers chosen by the publisher for reading all or part of my manuscript. The thoughtful comments of these experts and non-experts made this a much better book. I must also thank my copy editor Lyman Lyons for his careful editing of my manuscript and his many suggestions for improving the writing.

Finally, I would like to thank my wife Becky and children Ben and Nora, who put up with the many extra hours I spent working on this book.

APPENDIX A

BUILD YOUR OWN
POLYHEDRA AND SURFACES

A great way to learn about polyhedra and topological surfaces is to build them by hand. The pages that follow contain templates for making the five Platonic solids, the torus, the cylinder, the Möbius band, the Klein bottle, and the projective plane.

Here is some advice for making paper models:

1. Photocopy the templates from the book. Enlarge the copy to get larger models.
2. Copy the templates on heavy paper or card stock. Alternately, copy onto an adhesive sheet and stick this onto a piece of cardboard.
3. Carefully cut out the templates using a sharp knife and a straightedge.
4. Score the folds with a dull knife to guarantee nice, sharp creases.
5. Use the tabs if you want to glue the model together. Cut the tabs off if you intend to tape the model together.

Tetrahedron

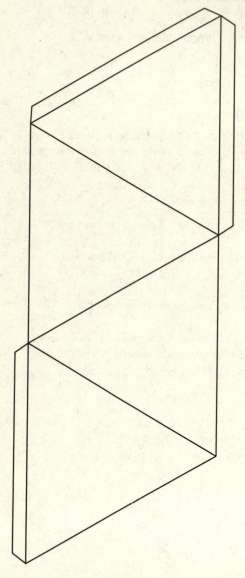

Cube

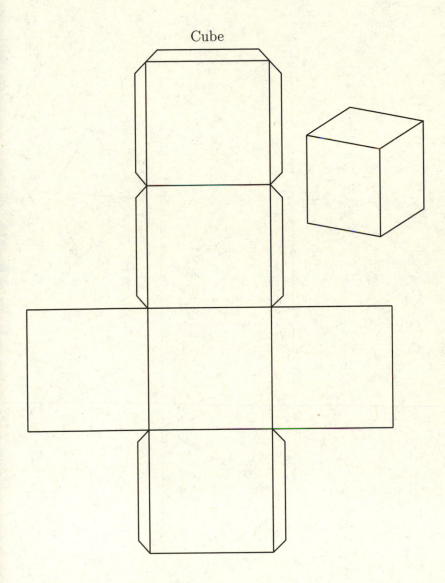

Octahedron

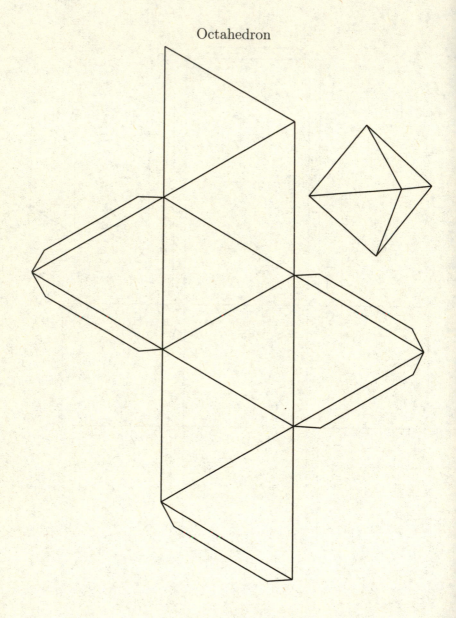

Icosahedron

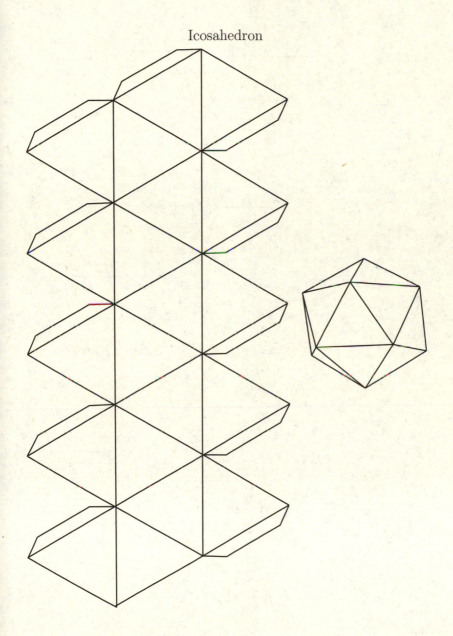

Dodecahedron

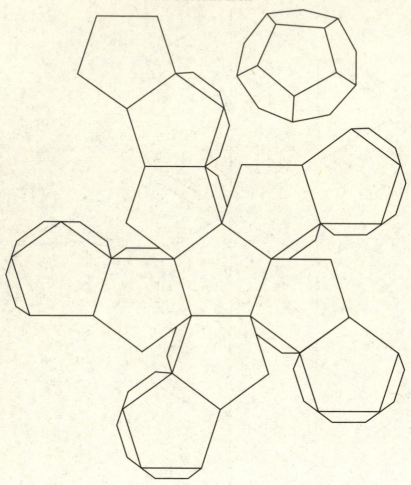

Torus

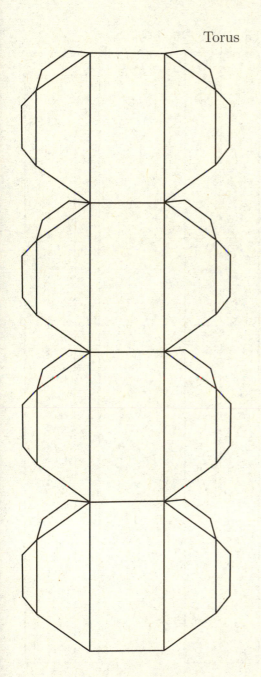

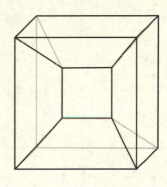

Möbius bands (or cylinders)

Klein bottle

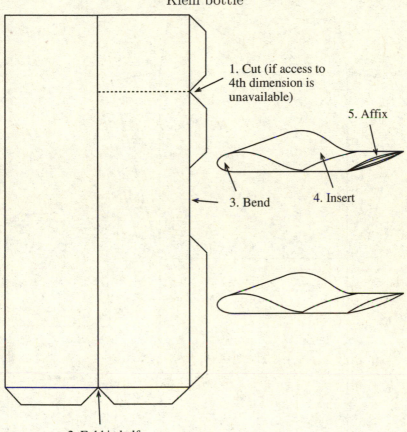

1. Cut (if access to 4th dimension is unavailable)

5. Affix

3. Bend

4. Insert

2. Fold in half

Projective plane

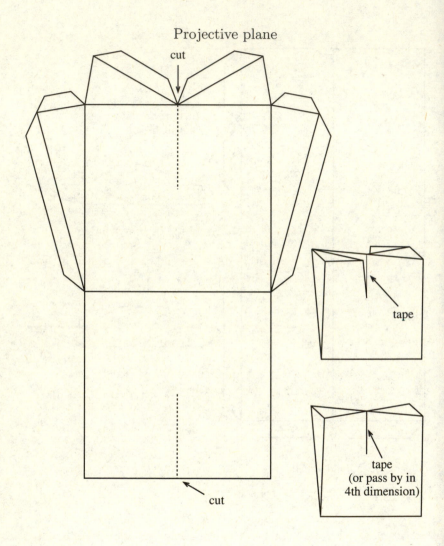

Appendix B

Recommended Readings

There is a complete bibliography at the end of this book. It contains all the sources that I used to prepare this book as well as many of the primary sources for the theorems you read about. I would like to highlight a few of these books and articles for the benefit of those readers who would like to learn more about the topics in this book. I will give the title and author. To find the complete bibliographic information, consult the references.

There are many good reference books for the history of mathematics. The first one I turn to is Carl Boyer and Uta Merzbach's *A History of Mathematics*. For biographies I recommend the eighteen-volume *Dictionary of Scientific Biography*, which can be found in the reference section of the library. It contains high-quality biographies written by experts in the field. For thoroughly entertaining, if not always historically accurate biographies, read Eric Temple Bell's 1937 classic *Men of Mathematics*. A very convenient resource is John O'Connor and Edmund Robertson's online MacTutor History of Mathematics Archive. This searchable Web-site contains numerous biographies as well as other valuable historical information.

There are many good resources on Euler, including a number of new ones that appeared in 2007, "the year of Euler." I recommend two new edited volumes, *The Genius of Euler: Reflections on His Life and Work*, edited by William Dunham, and *Leonhard Euler: Life, Work and Legacy*, edited by Robert Bradley and Edward Sandifer. Euler's collected works are available in the seventy-six-volume *Opera omnia*, and most are available online at Dominic Klyve and Lee Stemkoski's Euler Archive. A complete English translation of Euler's "Demonstratio nonnullarum insignium proprietatum quibus solida hedris planis inclusa sunt praedita" by Chris Francese and the author is available at the Euler Archive.

For more information on polyhedra, consult Peter Cromwell's *Poly-hedra*. Many of the topics in this book can be found there. In this wonderful book, Cromwell presents both the history and the theory of polyhedra.

The definitive discussion of the development of Euler's formula is Imre Lakatos's *Proofs and Refutations: The Logic of Mathematical Discovery*.

In this classic work, Lakatos, a philosopher of mathematics, uses Euler's formula and the many proofs, exceptions, and generalizations to present his view of mathematical discovery. The detailed footnotes were extremely useful to me as I researched this book.

To learn more about the history and content of Descartes' lost manuscript *The Elements of Solids,* read P. J. Federico's *Descartes on Polyhedra: A Study of the De Solidorum Elementis.* In addition to commentary, it contains reproductions of Leibniz's copies of Descartes' notes and English translations of them.

For topics in graph theory one cannot do better than the wonderful books and papers of Robin Wilson and his coauthors. For a history of graph theory, see Norman Biggs, Keith Lloyd, and Robin Wilson's *Graph Theory, 1736–1936.* It is a well-written resource complete with translations of many important articles. For a myth-busting account of the bridges of Königsberg problem, consult Brian Hopkins and Robin Wilson's "The Truth about Königsberg." And for a captivating account of the proof of the four color theorem, pick up Wilson's *Four Colors Suffice: How the Map Problem Was Solved.*

There are two excellent references available on the history of topology. One is the mammoth (1,000+ pages) *History of Topology* edited by I. M. James, and the other is the slightly shorter *A History of Algebraic and Differential Topology: 1900–1960* by Jean Dieudonné. These high-level texts were written for professional mathematicians.

Accessible works on the Euler number, combinatorial topology, geometry, and high-dimensional manifolds include D. M. Y. Sommerville's *An Introduction to the Geometry of n Dimensions,* David Hilbert and Stephan Cohn-Vossen's *Geometry and the Imagination,* Maurice Fréchet and Ky Fan's *Initiation to Combinatorial Topology,* and Jeffrey Weeks's *The Shape of Space.* For those who read French I recommend Jean-Claude Pont's *La topologie algèbrique des origines à Poincaré.*

To those who enjoy building topological surfaces out of paper and would like other activities, I recommend Stephen Barr's *Experiments in Topology.* Every mathematics enthusiast should seek out Martin Gardner's many marvelous books. They are full of fascinating mathematical gems. The Mathematical Association of America recently released fifteen of his books on one CD-ROM called *Martin Gardner's Mathematical Games: The Entire Collection of his Scientific American Columns.*

See Martin Aigner and Günter Ziegler's *Proofs from The Book* for a few other nice applications of Euler's formula. They also give an elementary proof of Cauchy's rigidity theorem for convex polyhedra. For an elegant and visual proof of the classification theorem for surfaces, see "Conway's ZIP proof" by George Francis and Jeffrey Weeks. Read Edwin Abbott's 1884 classic *Flatland: A Romance of Many Dimensions* to get a social satire together with an investigation of the notion of mathematical dimension, all in one novella.

For an introduction to knot theory, read Colin Adams's *The Knot Book: An Elementary Introduction to the Mathematical Theory of Knots*. It can be used as a textbook, but it reads like a pop math book.

Sylvia Nasar and David Gruber wrote an article for *The New Yorker* titled "Manifold Destiny: A Legendary Problem and the Battle Over Who Solved it." In it they detail the controversy surrounding the proof of the Poincaré conjecture and Thurston's geometrization conjecture.

Mathematicians voted Euler's polyhedron formula the second-most beautiful theorem in all of mathematics. We have seen a few other top-10 theorems as well: there are five regular polyhedra (#4), the Brouwer fixed point theorem (#6), $\sqrt{2}$ is irrational (#7), and the four color theorem (#9). To see the rest of the list, consult David Wells's article "Are These the Most Beautiful?"

Notes

Preface

1. Quoted in Schechter (1998), 155.

Introduction

1. Quoted in Machamer (1998).
2. Juškevič and Winter (1965), 333.

Chapter 1. Leonhard Euler and His Three "Great" Friends

1. Quoted in Dunham (1999), xiii.
2. Quoted in Youschkevitch (1971)
3. Riasanovsky (1993), 285.
4. Vucinich (1963), 69.
5. Quoted in Condorcet (1786)
6. Quoted in Eves (1969b), 48.
7. Quoted in Boyer and Merzbach (1991), 440.
8. Quoted in Cajori (1927)
9. Quoted in Calinger (1996)
10. Quoted in Cajori (1927)
11. Riasanovsky (1993), 248.
12. Quoted in Alexander (1989), 173.
13. Weil (1984)
14. Hartley (2003)
15. Hardy (1992), 70.
16. Vucinich (1963), 146–47.
17. Condorcet (1786)
18. Wells (1990)

Chapter 2. What Is a Polyhedron?

1. Hemingway (1932), 122.
2. Francese and Richeson (2007)
3. Poincaré (1913), 434.

CHAPTER 3. THE FIVE PERFECT BODIES

1. McEwan (1997), 20.
2. Plato (1972), 244.
3. Waterhouse (1972)
4. Ibid.

CHAPTER 4. THE PYTHAGOREAN BROTHERHOOD AND PLATO'S ATOMIC THEORY

1. Simmons (1992), 20.
2. Burkert (1972), 109.
3. Quoted in van der Waerden (1954), 94.
4. Quoted in Euclid (1926) vol. 3, 438.
5. Quoted in van der Waerden (1954), 165.
6. Taylor (1929), 5.
7. Boyer and Merzbach (1991), 84.
8. Allan (1975)
9. Plato (2000), 46.

CHAPTER 5. EUCLID AND HIS ELEMENTS

1. Russell (1967), 37–38.
2. Quoted in Bulmer-Thomas (1976)
3. Ibid.
4. van der Waerden (1954), 173.
5. Euler (1862)
6. Cauchy (1813a)
7. Connelly (1977)
8. Quoted in Bulmer-Thomas (1967), 195.

CHAPTER 6. KEPLER'S POLYHEDRAL UNIVERSE

1. Simmons (1992), 69.
2. Koestler (1963), 262.
3. Ibid, 252.
4. Kepler (1596), English translation in Kepler (1981)
5. Kepler (1596), quoted in Gingerich (1973)
6. Kepler (1981), 107.
7. Quoted in Martens (2000), 146.
8. Kepler (1938), English translation in Kepler (1997)
9. Quoted in Emmer (1993)

CHAPTER 7. EULER'S GEM

1. Bell (1987), 16.
2. Juškevič and Winter (1965), 333.

3. Euler (1758b)
4. Legendre (1794)
5. Juškevič and Winter (1965), 333.
6. Ibid., 334.
7. Euler (1758b)
8. Ibid.
9. Euler (1758a), English translation in Euler (1758c)
10. Legendre (1794)
11. Euler (1758a), English translation in Euler (1758c)
12. Lebesgue (1924)
13. Francese and Richeson (2007); Samelson (1996)

Chapter 8. Platonic Solids, Golf Balls, Fullerenes, and Geodesic Domes

1. Bell (1945), 211.
2. Poincaré (1913), 44.

Chapter 9. Scooped by Descartes?

1. Descartes (1965), 259.
2. Quoted in Bell (1937), 35.
3. Descartes (1965)
4. Kuhn (1970), 54.
5. Quoted in Federico (1982), 76.
6. Lebesgue (1924)
7. Kuhn (1970), 55.

Chapter 10. Legendre Gets It Right

1. Albers (1994)
2. Lohne (1972)
3. Quoted in Itard (1972)
4. Girard (1629)
5. Quoted in Itard (1972)
6. Poinsot (1810)
7. Ibid.

Chapter 11. A Stroll through Königsberg

1. Thoreau (1894), 419.
2. Quoted in Sachs, Stiebitz, and Wilson (1988)
3. Ibid.
4. Quoted in Hopkins and Wilson (2004)

5. Euler (1736), English translation in Biggs, Lloyd, and Wilson (1986), 3–8.
6. Ball (1892)
7. Hierholzer (1873)
8. Barabási (2002), 12.
9. Listing (1847)
10. Terquem (1849)

Chapter 12. Cauchy's Flattened Polyhedra

1. Abel (1881), 259.
2. Freudenthal (1971)
3. Ibid.
4. Simmons (1992), 186.
5. Cauchy (1813a)
6. Lhuilier (1813)
7. Ibid.
8. Hadamard (1907)
9. Listing (1861–62); Jordan (1866b)

Chapter 13. Planar Graphs, Geoboards, and Brussels Sprouts

1. Hankel (1884)
2. Hardy (1992), 94.
3. Pick (1899)
4. DeTemple (1989)
5. Gardner (1975a), 8.
6. Applegate, Jacobson, and Sleator (1991)

Chapter 14. It's a Colorful World

1. Twain (1894), 42–43.
2. May (1965)
3. Graves (1889), 423.
4. Ibid.
5. Cayley (1878)
6. Quoted in Dudley (1992)
7. Gardner (1975b); Gardner (1988)
8. Baltzer (1885), quoted in Coxeter (1959)
9. Ibid.
10. Kempe (1879)
11. Quoted in Wilson (2002), 119.
12. Gardner (1995)
13. Appel and Haken (1977); Appel, Haken, and Koch (1977)
14. Hales (2005)

Chapter 15. New Problems and New Proofs

1. Quoted in Federico (1982), 71.
2. Sommerville (1958), 143–44.
3. de Jonquières (1890)
4. Speziali (1973)
5. Pont (1974), 24.
6. Lhuilier (1813)
7. Hessel (1832)
8. Poinsot (1810)
9. Cauchy (1813a)
10. Lhuilier (1813)
11. Steiner (1826)
12. von Staudt (1847), 18–23.
13. Hoppe (1879)

Chapter 16. Rubber Sheets, Hollow Doughnuts, and Crazy Bottles

1. Listing (1847)
2. Tait (1883)
3. Lefschetz (1970)
4. In an interview in Maurer (1983)
5. Klein (1882/83)
6. Brahana (1921)
7. Clarke (2000)
8. Gardner (1990)
9. Gardner (1956)
10. Listing (1861–62)
11. Möbius (1865)
12. In the introduction of Abbott (2005), xxix.
13. Klein (1882)

Chapter 17. Are They the Same, or Are They Different?

1. Poincaré (1895)
2. Möbius (1863)
3. Radó (1925)
4. Papakyriakopoulos (1943)
5. Quoted in Freudenthal (1975)
6. Riemann (1851); Riemann (1857)
7. Möbius (1863)
8. Jordan (1866a)
9. Dyck (1888)
10. Dehn and Heegaard (1907)
11. Francis and Weeks (1999)
12. Ibid.

CHAPTER 18. A KNOTTY PROBLEM

1. Shakespeare (2002), 82.
2. Vandermonde (1771)
3. Gauss (1877)
4. Listing (1847)
5. Seifert (1934)
6. Sequence number A002864 in Sloane (2007)
7. Crowell (1959)
8. Sequence number A002863 in Sloane (2007)
9. Kauffman (1987b); Murasugi (1987); Thistlethwaite (1987)

CHAPTER 19. COMBING THE HAIR ON A COCONUT

1. Frost (2002), 308.
2. Brouwer (1912)
3. Quoted in Dieudonné (1975)
4. Dieudonné (1975)
5. Poincaré (1881)
6. Poincaré (1885)
7. Brouwer (1912)
8. Beno Eckmann, quoted in Frei and Stammbach (1999)
9. Hopf (1925); Hopf (1926a); Hopf (1926b)
10. Morse (1929)
11. Thurston (1997)
12. Brouwer (1909)
13. Brouwer (1912)

CHAPTER 20. WHEN TOPOLOGY CONTROLS GEOMETRY

1. Shakespeare (1992), 36.
2. Pólya (1954), 57–58.
3. Hopf (1935)
4. Quoted in Federico (1982), 43.
5. Euler (1758b); Euler (1758a)

CHAPTER 21. THE TOPOLOGY OF CURVY SURFACES

1. Bell (1937), 254
2. Euler (1760)
3. See Hayes (2006) for a discussion of this story.
4. Quoted in Simmons (1992), 177.
5. Gauss (1828); English translation and commentary in Dombrowski (1979)
6. Bonnet (1848)

7. Blaschke (1921)
8. Dyck (1888)

CHAPTER 22. NAVIGATING IN *n* DIMENSIONS

1. Scholz (1999)
2. Brouwer (1911)
3. Cauchy (1813a)
4. Schläfli (1901)
5. Breitenberger (1999)
6. Listing (1847), Listing (1861–62)
7. Tait (1884)
8. Riemann (1851)

CHAPTER 23. HENRI POINCARÉ AND THE ASCENDANCE OF TOPOLOGY

1. Hardy (1992), 85.
2. Quoted in Dieudonné (1975)
3. Poincaré (1895)
4. Poincaré (1899); Poincaré (1900); Poincaré (1902a); Poincaré (1902b); Poincaré (1904)
5. Dieudonné (1989), 17.
6. Heinrich Tietze (1880–1964), quoted in James (2001)
7. Poincaré (1895)
8. Poincaré (1895), quoted in Sarkaria (1999).
9. Quoted in James (1999)
10. Ibid.

EPILOGUE: THE MILLION-DOLLAR QUESTION

1. Russell (1957)
2. Poincaré (1900)
3. Poincaré (1904)
4. Ibid.
5. Taubes (1987)
6. Smale (1961)
7. Smale (1990)
8. Ibid.
9. Stallings (1962); Stallings (1960); Zeeman (1961); Zeeman (1962)
10. Freedman (1982)
11. Smale (1998)
12. Thurston (1982)
13. Hamilton (1982)
14. Perelman (2002); Perelman (2003b); Perelman (2003a)

15. Cao and Zhu (2006a); Cao and Zhu (2006b); Kleiner and Lott (2006); Morgan and Tian (2006)
16. Mackenzie (2006)
17. See Nasar and Gruber (2006) for more details.
18. Nasar and Gruber (2006)
19. Quoted in Nasar and Gruber (2006)
20. Poincaré (1913), 366.

REFERENCES

Abbott, E. A. (2005). *Flatland: A romance of many dimensions.* Princeton, NJ: Princeton University Press. With an introduction by Thomas Banchoff.

Abel, N. H. (1881). *Oeuvres completes de Niels Henrik Abel,* vol. 2. Christiania, Norway: Imprimerie De Grondahl & Son.

Adams, C. C. (1994). *The knot book: An elementary introduction to the mathematical theory of knots.* New York: W. H. Freeman.

Aigner, M., and G. M. Ziegler (2001). *Proofs from The Book* (2nd ed.). Including illustrations by Karl H. Hofmann. Berlin: Springer-Verlag.

Albers, D. J. (1994). Freeman Dyson: Mathematician, physicist, and writer. *The College Mathematics Journal* 25(1), January, 2–21.

Alexander, J. T. (1989). *Catherine the Great: Life and legend.* New York: Oxford University Press.

Allan, D. J. (1975). Plato. In C. C. Gillispie (ed.), *Dictionary of scientific biography.* Vol. 11, 22–31. New York: Charles Scribner's Sons.

Andrews, P. (1988). The classification of surfaces. *Amer. Math. Monthly* 95(9), 861–67.

Appel, K., and W. Haken (1977). Every planar map is four colorable. I. Discharging. *Illinois J. Math.* 21(3), 429–90.

Appel, K., W. Haken, and J. Koch (1977). Every planar map is four colorable. II. Reducibility. *Illinois J. Math.* 21(3), 491–567.

Applegate, D., G. Jacobson, and D. Sleator (1991). *Computer analysis of sprouts.* Technical Report CMU-CS-91-144, Carnegie Mellon University.

Asimov, I. (1965). *A short history of chemistry: An introduction to the ideas and concepts of chemistry.* Science Study Series. Garden City, NY: Anchor Books, Doubleday.

Ball, W. W. R. (1892). *Mathematical recreations and problems of past and present times.* London: MacMillan.

Baltzer, R. (1885). Eine Erinnerung an Möbius und seinen Freund Weiske. *Ber. Verh. K. Sächs. Ges. Wiss. Leipzig* 37, 1–6.

Barabási, A.-L. (2002). *Linked: How everything is connected to everything else and what it means.* Cambridge, MA: Perseus.

Barnette, D. (1983). *Map coloring, polyhedra, and the four-color problem.* Washington DC: Mathematical Association of America.

Barr, S. (1964). *Experiments in topology.* New York: Dover.

Baxter, M. (1990). Unfair games. *Ureka: The Journal of the Archimedeans* 50, 60–68.

Becker, J. C., and D. H. Gottlieb (1999). A history of duality in algebraic topology. In I. M. James (ed.), *History of topology,* 725–45. Amsterdam: North-Holland.

Bell, E. T. (1937). *Men of Mathematics.* New York: Simon and Schuster.

———. (1945). *The development of mathematics.* New York: McGraw-Hill.

———. (1987). *Mathematics: Queen and servant of science.* MAA Spectrum series. Washington DC: Mathematical Association of America.

Biggs, N. (1993). The development of topology. In J. Fauvel, R. Flood, and R. Wilson (eds.), *Möbius and his band: Mathematics and astronomy in nineteenth-century*

Germany, 105–19. New York: The Clarendon Press, Oxford University Press.

Biggs, N. L., E. K. Lloyd, and R. J. Wilson (1986). *Graph theory, 1736–1936*. Oxford: Clarendon Press.

Blaschke, W. (1921). *Vorlesungen über Differentialgeometrie*. Berlin-Heidelberg: Springer-Verlag.

Bonnet, O. (1848). Mémoire sur la théorie générale des surfaces. *J. Éc. Polytech.* 19, 1–146.

Boyer, C. B. (1951). The foremost textbook of modern times. *Amer. Math. Monthly* 58, April, 223–26.

Boyer, C. B., and U. Merzbach (1991). *A history of mathematics* (2nd ed.). New York: John Wiley & Sons.

Boyle, R. (1937). *The sceptical chymist, with an introduction by M. M. Pattison Muir*. Everyman's Library. London: J. M. Dent and Sons.

Bradley, R., and E. Sandifer (eds.) (2007). *Leonhard Euler: Life, work and legacy*. Vol. 5 of *Studies in the history and philosophy of mathematics*. Amsterdam: Elsevier.

Brahana, H. R. (1921). Systems of circuits on two-dimensional manifolds. *Ann. of Math.* (2) 23(2), 144–68.

Breitenberger, E. (1999). Johann Benedict Listing. In I. M. James (ed.), *History of topology*, 909–24. Amsterdam: North-Holland.

Brewster, D. (1833). The life of Euler. In *Letters of Euler on different subjects in natural philosophy addressed to a German princess*, 15–28. New York: J. and J. Harper.

Brisson, L., and F. W. Meyerstein (1995). *Inventing the universe: Plato's "Timaeus," the big bang, and the problem of scientific knowledge*. Albany, NY: State University of New York Press.

Brouwer, L. E. J. (1909). On continuous one-to-one transformations of surfaces to themselves. *Proc. Kon. Nederl. Akad. Wetensch. Ser. A* 11, 788–98.

———. (1911). Beweis der Invarianz der Dimensionzahl. *Mathematische Annalen* 69, 169–75.

———. (1912). Über Abbildung von Mannigfaltigkeiten. *Mathematische Annalen* 71(1), 97–115.

Bulmer-Thomas, I. (1967). *Selections illustrating the history of Greek mathematics with an English translation by Ivor Thomas*, Vol. 2 of *Loeb Classical Library*. Cambridge, MA: Harvard University Press.

———. (1971). Euclid. In C. C. Gillispie (ed.), *Dictionary of scientific biography*. Vol. 4, 414–37. New York: Charles Scribner's Sons.

———. (1976). Theaetetus. In C. C. Gillispie (ed.), *Dictionary of scientific biography*. Vol. 13, 301–7. New York: Charles Scribner's Sons.

Burau, W. (1976). Staudt, Karl Georg Christian von. In C. C. Gillispie (ed.), *Dictionary of scientific biography*. Vol. 8, 4–6. New York: Charles Scribner's Sons.

Burckhardt, J. J. (1983). Leonhard Euler, 1707–1783. *Math. Mag.* 56(5), 262–73.

Burde, G., and H. Zieschang (1999). Development of the concept of a complex. In *History of topology*, 103–10. Amsterdam: North-Holland.

Burke, J. G. (1972). Hessel, Johann Friedrich Christian. In C. C. Gillispie (ed.), *Dictionary of scientific biography*. Vol. 6, 358–59. New York: Charles Scribner's Sons.

Burkert, W. (1972). *Lore and science in ancient Pythagoreanism*. Cambridge, MA: Harvard University Press.

Cajori, F. (1927). Frederick the Great on mathematics and mathematicians. *Amer. Math. Monthly* 34, 122–30.

Calinger, R. (1968). Frederick the Great and the Berlin Academy of Sciences (1740–1766). *Annals of Science* 24(3), 239–49.

———. (1996). Leonhard Euler: The first St. Petersburg years (1727–1741). *Historia Math.* 23(2), 121–66.

Cao, H.-D., and X.-P. Zhu (2006a). A complete proof of the Poincaré and geometrization conjectures—application of the Hamilton-Perelman theory of the Ricci flow. *Asian Journal of Mathematics* 10(2), June, 165–492.

———. (2006b). Erratum to "A complete proof of the Poincaré and geometrization conjectures—application of the Hamilton-Perelman theory of the Ricci flow." *Asian Journal of Mathematics* 10(4), December, 663–64.

Carruccio, E. (1970). Betti, Enrico. In C. C. Gillispie (ed.), *Dictionary of scientific biography*. Vol. 2, 104–6. New York: Charles Scribner's Sons.

Casselman, B. (2004). *Mathematical illustrations: A manual of geometry and Post-Script*. New York: Cambridge University Press.

Cauchy, A. L. (1813a). Recherches sur les polyèdres. *Journal de l'École Polytechnique* 9, 68–86.

———. (1813b). Sur les polygones et les polyèdres. *Journal de l'École Polytechnique* 9, 87–98.

Cayley, A. (1861). On partitions of a close. *The London, Edinburgh, and Dublin Philosophical Magazine and Journal of Science*, 4th Series 21, 424–28.

———. (1873). On Listing's theorem. *Messenger of Mathematics II*, 81–89.

———. (1878). On the coloring of maps. *Proceedings of the London Mathematical Society IX*, 14.

Chambers, M., R. Grew, D. Herlihy, T. Rabb, and I. Woloch (1987). *The Western experience, Vol. 2: The early modern period* (4th ed.). New York: Alfred A. Knopf.

Clarke, A. C. (2000). The wall of darkness. In *The Collected Stories of Arthur C. Clarke*, 104–18. New York: Tom Doherty Associates.

Collingwood, S. D. (1898). *The life and letters of Lewis Caroll (Rev. C. L. Dodgson)*. London: T. Fisher Unwin.

Condorcet, M. J. (1786). Eulogy to Mr. Euler (translated by John S. D. Glaus, Euler Society, 2005). In *History of the Royal Academy of Sciences 1783*, 37–68. Paris: Imprimerie Royale.

Connelly, R. (1977). A counterexample to the rigidity conjecture for polyhedra. *Inst. Hautes Études Sci. Publ. Math.* 47, 333–38.

Conway, J., P. Doyle, J. Gilman, and B. Thurston (1994). Geometry and the imagination. http://www.geom.uiuc.edu/docs/education/institute91.

Coolidge, J. L. (1963). *A history of geometrical methods*. New York: Dover.

Copper, M. (1993). Graph theory and the game of sprouts. *Amer. Math. Monthly* 100(5), 478–82.

Coxeter, H. S. M. (1959). The four-color map problem, 1840–1890. *Mathematics Teacher* 52, 283–89.

———. (1988). Regular and semiregular polyhedra. In M. Senechal and G. Fleck (eds.), *Shaping space: A polyhedral approach*, proceedings of 1984 conference held in Northampton, MA, 67–79. Boston, Design Science Collection, Birkhäuser Boston.

Crombie A. C., (1971). Descartes, Réne du Perron. In Charles C. Gillispie (ed.), *Dictionary of scientific biography*, vol. 4, 51–55. New York: Charles Scribner's Sons.

Cromwell, P. R. (1995). Kepler's work on polyhedra. *Math. Intelligencer* 17(3), 23–33.

———. (1997). *Polyhedra*. Cambridge: Cambridge University Press.

Crowe, M. J. (1974). Möbius, August Ferdinand. In C. C. Gillispie (ed.), *Dictionary of scientific biography*. Vol. 9, 429–31. New York: Charles Scribner's Sons.

Crowell, R. (1959). Genus of alternating link types. *Ann. of Math.* (2) 69, 258–75.

D'Alarcao, H., and T. E. Moore (1976–77). Euler's formula and a game of Conway's. *Journal of Recreational Mathematics* 9(4), 149–251.

de Jonquières, E. (1890). Note sur un point fondamental de la théorie des polyèdres. *Comptes Rendus des Séances de l'Académie des Sciences* 110, 110–15.

Dehn, M., and P. Heegaard (1907). Analysis situs. In *Enzyklopädie Mathematische Wissenschaft*. Vol. 3, 153–220. Leipzig: Teubner.

Descartes, R. (1965). *Discourse on method, Optics, geometry, and meteorology*. Translated with an introduction by Paul J. Olscamp. Indianapolis: Bobbs-Merrill.

DeTemple, D. (1989). Pick's theorem: A retrospective. *Mathematics Notes From Washington State University* 32(3 and 4), November, 1–4.

DeTemple, D., and J. M. Robertson (1974). The equivalence of Euler's and Pick's theorems. *Math. Teacher* 67(3), 222–26.

Dieudonné, J. (1975). Poincaré, Henri Jules. In C. C. Gillispie (ed.), *Dictionary of scientific biography*. Vol. 11, 51–61. New York: Charles Scribner's Sons.

———. (1989). *A history of algebraic and differential topology: 1900–1960*. Boston: Birkhäuser Boston.

Dombrowski, P. (1979). *150 years after Gauss' "Disquisitiones generales circa superficies curvas."* With the original text of Gauss. Vol. 62 of *Astérisque*. Paris: Société Mathématique de France.

Doyle, P. H., and D. A. Moran (1968). A short proof that compact 2-manifolds can be triangulated. *Invent. Math.* 5, 160–62.

Dudley, U. (1992). *Mathematical cranks*. MAA Spectrum series. Washington DC: Mathematical Association of America.

Dunham, W. (1990). *Journey through genius: The great theorems of mathematics*. New York: John Wiley & Sons.

———. (1999). *Euler: The master of us all*. Vol. 22 of *The Dolciani Mathematical Expositions*. Washington DC: Mathematical Association of America.

———. (ed.) (2007). *The genius of Euler: Reflections on his life and work*. MAA Spectrum series. Washington DC: Mathematical Association of America.

Dyck, W. (1888). Beiträge zur Analysis situs. *Math. Ann. XXXII*, 457–512.

Emmer, M. (1993). Art and mathematics: The Platonic solids. In M. Emmer (ed.), *The visual mind*, Leonardo Book Ser., 215–20. Cambridge, MA: MIT Press.

Epple, M. (1999). Geometric aspects in the development of knot theory. In *History of topology*, 301–57. Amsterdam: North-Holland.

Euclid (1926). *The thirteen books of Euclid's Elements translated from the text of Heiberg. Vol. I: Introduction and Books I, II. Vol. II: Books III–IX. Vol. III: Books X–XIII and Appendix*. Translated with introduction and commentary by Thomas L. Heath, 2nd ed. Cambridge: Cambridge University Press.

Euler, L. (1736). Solutio problematis ad geometriam situs pertinentis. *Commentarii Academiae Scientiarum Imperialis Petropolitanae* 8, 128–40. Also in *Opera Omnia* series 1, vol. 7, 1–10.

———. (1758a). Demonstratio nonnullarum insignium proprietatum quibus solida hedris planis inclusa sunt praedita. *Novi Commentarii Academiae Scientiarum Petropolitanae* 4, 94–108. Also in *Opera Omnia* series 1, vol. 26, 94–108.

———. (1758b). Elementa doctrinae solidorum. *Novi Commentarii Academiae Scientiarum Petropolitanae* 4, 72–93. Also in *Opera Omnia* series 1, vol. 26, 71–93.

———. (1758c). Proof of some notable properties with which solids enclosed by plane faces are endowed. Translated by Chris Francese and David Richeson. http://www.eulerarchive.org.

———. (1760). Recherches sur la courbure des surfaces. *Mém. Acad. Sci. Berlin* 16, 119–43.

———. (1862). Continuatio fragmentorum ex adversariis mathematicis deprompto-rum. In P. H. Fuss and N. Fuss (eds.), *Opera postuma mathematica et physica*, Vol. 1, 487–518. St. Petersburg: St. Petersburg Academy of Sciences.

———. (1911–). *Opera Omnia*. Basel: Birkhäuser.

Euler Archive, http://www.eulerarchive.org. Created by D. Klyve and L. Stemkoski.

Eves, H. (1990). *An introduction to the history of mathematics*. 6th ed. Pacific Grove, CA: Brooks/Cole.

———. (1969a). *In mathematical circles, quadrants I and II*. Boston: Prindle, Weber & Schmidt.

———. (1969b). *In mathematical circles, quadrants III and IV*. Boston: Prindle, Weber & Schmidt.

Federico, P. J. (1982). *Descartes on polyhedra: A study of the De Solidorum Elementis*. Vol. 4 of *Sources in the history of mathematics and physical sciences*. New York: Springer-Verlag.

Field, J. V. (1979). Kepler's star polyhedra. *Vistas Astronom.* 23(2), 109–41.

———. (1988). *Kepler's geometrical cosmology*. Chicago: University of Chicago Press.

Francese, C., and D. Richeson (2007). The flaw in Euler's proof of his polyhedral formula. *Amer. Math. Monthly* 114, April, 286–96.

Francis, G. K. (1987). *A topological picturebook*. New York: Springer-Verlag.

Francis, G. K., and J. R. Weeks (1999). Conway's ZIP proof. *Amer. Math. Monthly* 106(5), 393–99.

Fréchet, M., and K. Fan (1967). *Initiation to combinatorial topology*. Translated from the French, with some notes, by H. W. Eves. Boston: Prindle, Weber & Schmidt.

Freedman, M. H. (1982). The topology of four-dimensional manifolds. *J. Differential Geom.* 17(3), 357–453.

Frei, G. and U. Stammbach (1999). Heinz Hopf. In I. M. James (ed.), *History of topology*, 991–1008. Amsterdam: North-Holland.

Freudenthal, H. (1971). Cauchy, Augustin-Lous. In C. C. Gillispie (ed.), *Dictionary of scientific biography*. Vol. 3, 131–48. New York: Charles Scribner's Sons.

———. (1972). Hopf, Heinz. In C. C. Gillispie (ed.), *Dictionary of scientific biography*. Vol. 6, 496–97. New York: Charles Scribner's Sons.

———. (1975). Riemann, Georg Friedrich Bernhard. In C. C. Gillispie (ed.), *Dictionary of scientific biography*. Vol. 11, 447–56. New York: Charles Scribner's Sons.

Fritsch, R. and G. Fritsch (1998). *The four-color theorem: History, topological foundations, and idea of proof.* Translated from the 1994 German original by Julie Peschke. New York: Springer-Verlag.

Frost, R. (2002). *The poetry of Robert Frost: The collected poems.* New York: Henry Holt.

Funkenbusch, W. W. (1974). Classroom notes: From Euler's formula to Pick's formula using an edge theorem. *Amer. Math. Monthly* 81(6), 647–48.

Gardner, M. (1956). The Afghan bands. In *Mathematics, Magic and Mystery*, 70–73. New York: Dover.

———. (1975a). *Mathematical Carnival.* New York: Knopf.

———. (1975b). Mathematical games: Six sensational discoveries that somehow or another have escaped public attention. *Scientific American* 232, April, 127–31.

———. (1988). Six sensational discoveries. In *Time Travel and Other Mathematical Bewilderments*, 125–38. New York: W. H. Freeman.

———. (1990). Möbius bands. In *Mathematical Magic Show*, 123–36. Washington DC: Mathematical Association of America.

———. (1995). The four-color map theorem. In *New Mathematical Diversions*, 113–23. Washington DC: Mathematical Association of America.

———. (2005). *Martin Gardner's mathematical games: The entire collection of his Scientific American columns.* MAA Spectrum series (on CD). Mathematical Association of America.

Gaskell, R. W., M. S. Klamkin, and P. Watson (1976). Triangulations and Pick's theorem. *Math. Mag.* 49(1), 35–37.

Gaukroger, S. (1995). *Descartes: An intellectual biography.* Oxford: Clarendon Press.

Gauss, C. F. (1828). *Disquisitiones generales circa superficies curvas.* Göttingen: Dieterich.

———. (1877). Zur mathematischen Theorie der electrodynamischen Wirkungen. In *Werke*, vol. 5. Königlichen Gesellschaft der Wissenchaften zu Göttingen.

Gillispie, C. C. (ed.) (1970–1990). *Dictionary of Scientific Biography*, Volumes 1–16. New York: Charles Scribner's Sons.

Gingerich, O. (1973). Kepler, Johannes. In C. C. Gillispie (ed.), *Dictionary of scientific biography.* Vol. 7, 289–312. New York: Charles Scribner's Sons.

Girard, A. (1629). *Invention nouvelle en algèbre.* Amsterdam: Guillaume Jansson Blaeuw.

Gluck, H. (1975). Almost all simply connected closed surfaces are rigid. In *Geometric topology (Proc. Conf., Park City, Utah, 1974)*, 225–39. Lecture Notes in Math., Vol. 438. Berlin: Springer.

Gorman, P. (1979). *Pythagoras: A life.* London: Routledge & Kegan Paul.

Gottlieb, D. H. (1996). All the way with Gauss-Bonnet and the sociology of mathematics. *Amer. Math. Monthly* 103(6), 457–69.

Graham, L. R. (1993). *Science in Russia and the Soviet Union: A short history.* Cambridge: Cambridge University Press.

Graves, R. P. (1889). *Life of Sir William Rowan Hamilton*, vol. 3. Dublin: Hodges, Figgis, & Co.

Grünbaum, B. (2003). *Convex polytopes* (2nd ed.). Vol. 221 of *Graduate texts in mathematics.* Prepared and with a preface by Volker Kaibel, Victor Klee, and Günter M. Ziegler. New York: Springer-Verlag.

Grünbaum, B., and G. C. Shephard (1993). Pick's theorem. *Amer. Math. Monthly* 100(2), 150–61.

Guillemin, V., and A. Pollack (1974). *Differential topology*. Englewood Cliffs, NJ: Prentice-Hall.

Guthrie, W. (1962). *A history of Greek philosophy: The earlier presocratics and the Pythagoreans*, vol. 1. Cambridge: Cambridge University Press.

Hadamard, J. (1907). Erreurs de mathematicians. *L'intermédiaire des mathématiciens* 14, 31.

Hales, T. C. (2005). A proof of the Kepler conjecture. *Ann. of Math.* (2) 162(3), 1065–185.

Hamilton, R. S. (1982). Three-manifolds with positive Ricci curvature. *J. Differential Geom.* 17(2), 255–306.

Hankel, H. (1884). *Die Entwicklung der Mathematik in den letzten Jahrhunderten.* Tübingen: Verlag und Druck von Franz.

Harcave, S. (1964). *Russia: A history* (5th ed.). Philadelphia: J. B. Lippincott.

Hardy, G. H. (1992). *A mathematician's apology*. Cambridge: Cambridge University Press.

Hartley, J. M. (2003). Governing the city: St. Petersburg and Catherine II's reforms. In A. Cross (ed.), *St. Petersburg: 1703–1825*, 99–118. London: Palgrave Macmillan.

Hatcher, A. (2002). *Algebraic topology*. Cambridge: Cambridge University Press.

Hayes, B. (2006). Gauss's day of reckoning: A famous story about the boy wonder of mathematics has taken on a life of its own. *American Scientist* 94(3), 200–5.

Heath, T. (1981). *A history of Greek mathematics: From Aristarchus to Diophantus*, vol. II. Corrected reprint of the 1921 original. New York: Dover.

Hemingway, E. (1960). *Death in the afternoon*. New York: Charles Scribner & Sons.

Hessel, J. F. (1832). Nachtrag zu dem Eulerschen Lehrsatz von Polyëdern. *Journal für die Reine und Angewandte mathematik* 8, 13–20.

Hierholzer, C. (1873). Über die mglichkeit, einen linienzug ohne wiederholung und ohne unterbrechnung zu umfahren. *Mathematische Annalen* 6, 30–32.

Hilbert, D., and S. Cohn-Vossen (1952). *Geometry and the imagination*. Translated by P. Neményi. New York: Chelsea Publishing Company.

Hollingdale, S. (1989). *Makers of mathematics*. London: Penguin Books.

Hopf, H. (1925). Über die curvatura integra geschlossener Hyperflächen. *Math. Ann.* 95, 340–67.

———. (1926a). Abbildungsklassen *n*-dimensionaler Mannigfaltigkeiten. *Math. Ann.* 96, 209–24.

———. (1926b). Vectorfelder in *n*-dimensionalen Mannigfaltigkeiten. *Math. Ann.* 96, 225–49.

———. (1935). Über die Drehung der Tangenten und Sehnen ebener Kurven. *Comp. Math.* 2, 50–62.

Hopkins, B., and R. Wilson (2004). The truth about Königsburg. *Col. Math. J.* 35(3), May, 198–207.

Hoppe, R. (1879). Ergänzung des Eulerschen Satzes von den Polyedern. *Archiv der Mathematik und Physik* 63, 100–103.

Itard, J. (1972). Girard, Albert. In C. C. Gillispie (ed.), *Dictionary of scientific biography*. Vol. 5, 408–10. New York: Charles Scribner's Sons.

———. (1973). Legendre, Adrien-Marie. In C. C. Gillispie (ed.), *Dictionary of scientific biography*. Vol. 8, 135–43. New York: Charles Scribner's Sons.

Jackson, A. (2006). Conjectures no more? Consensus forming on the proof of the Poincaré and geometrization conjectures. *Notices Amer. Math. Soc.* 53(8), 897–901.

James, I. M. (1999). From combinatorial topology to algebraic topology. In *History of topology*, 561–73. Amsterdam: North-Holland.

———. (2001). Combinatorial topology versus point-set topology. In *Handbook of the history of general topology*, Vol. 3 of *History of Topology Series*, 809–34. Dordrecht: Kluwer Acad. Publ.

Jones, P. S. (1994). Irrationals or incommensurables I: Their discovery and a "logical scandal." In F. J. Swetz (ed.), *From five fingers to infinity: A journey through the history of mathematics*, 172–75. Chicago: Open Court.

Jordan, C. (1866a). Des contours tracés sur les surfaces. *J. Math. Pures Appl.* (2) 11, 110–30. See also *Oeuvres*, vol. 4, 91–112, Paris: Gauthier-Villars et cie, 1964.

———. (1866b). Recherches sur les polyèdres. *Comptes rendus des Séances de l'Académie des Sciences* 62, 1339–41.

Juškevič, A. P., and E. Winter (1965). *Leonhard Euler und Christian Goldbach: Briefwechsel 1729–1764*. Berlin: Akademie-Verlag.

Katz, V. J. (1993). *A history of mathematics: An introduction*. New York: Harper Collins College Publishers.

Kauffman, L. H. (1987a). *On knots*, vol. 115 of *Annals of mathematics studies*. Princeton, NJ: Princeton University Press.

———. (1987b). State models and the Jones polynomial. *Topology* 26(3), 395–407.

Kempe, A. B. (1879). On the geographical problem of four colors. *American Journal of Mathematics II*, 193–204.

Kepler, J. (1596). *Mysterium cosmographicum*. Tübingen.

———. (1938). Harmonice mundi. In M. Caspar (ed.), *Johannes Kepler Gesammelte Werke*, vol. 6. Munich: Beck.

———. (1981). *Mysterium cosmographicum: The secret of the universe*. Translation by A. M. Duncan, introduction and commentary by E. J. Aiton, with a preface by I. Bernard Cohen. New York: Abaris Books.

———. (1997). *The harmony of the world*, vol. 209 of *Memoirs of the American Philosophical Society*. Translated from the Latin and with an introduction and notes by E. J. Aiton, A. M. Duncan, and J. V. Field, with a preface by Duncan and Field. Philadelphia: American Philosophical Society.

Klein, F. (1882). *Über Riemanns Theorie der algebraischen Functionen und ihrer Integrale*. Leipzig: B. G. Teubner.

———. (1882/83). Neue Beiträge zur Riemannschen Funktionentheorie. *Math. Ann.* 21, 141–218.

Kleiner, B., and J. Lott (2006). Notes on Perelman's papers. http://www.math.lsa.umich.edu/~lott/ricciflow/perelman.html.

Kline, M. (1972). *Mathematical thought from ancient to modern times*. New York: Oxford University Press.

Knobloch, E. (1998). Mathematics at the Prussian Academy of Sciences 1700–1810. In H. G. W. Begehr, H. Koch, J. Kramer, N. Schnappacher, and E.-J. Thiele (eds.), *Mathematics in Berlin*, 1–8. Berlin: Birkhäuser.

Koestler, A. (1963). *The sleepwalkers: A history of man's changing vision of the universe.* New York: The Universal Library.

Kuhn, T. S. (1970). *The structure of scientific revolutions* (2nd ed.). Chicago: University of Chicago Press.

Lakatos, I. (1976). *Proofs and refutations: The logic of mathematical discovery.* Cambridge: Cambridge University Press.

Lebesgue, H. (1924). Remarques sur les deux premières démonstrations du théoréme d'Euler relatif aux polyèdres. *Bulletin de la Société Mathématique de France* 52, 315–36.

Lefschetz, S. (1970). The early development of algebraic topology. *Bol. Soc. Brasil. Mat.* 1(1), 1–48.

Legendre, A. M. (1794). *Éléments de géométrie.* Paris: Firmin Didot.

Lhuilier, S. (1811). Démonstration immédéate d'un théorème fondamental d'Euler sur les polyèdres. *Mémoires de l'Académie Impériale des Sciences de Saint-Pétersbourg* 4, 271–301.

———. (1813). Mémoire sur la polyédrométrie. *Annales de Mathématiques* 3, 169–89.

Listing, J. B. (1847). Vorstudien zur Topologie. *Göttinger studien (Abtheilung 1)* 1, 811–75.

———. (1861–62). Der Census räumlicher Complexe oder Verallgemeinerung des Euler'-schen Satzes von den Polyedern. *Abhandlungen der königlichen Gesellschaft der Wissenschaften zu Göttingen* 10, 97–182.

Liu, A. C. F. (1979). Lattice points and Pick's theorem. *Math. Mag.* 52(4), 232–35.

Livingston, C. (1993). *Knot theory,* vol. 24 of *Carus Mathematical Monographs.* Washington DC: Mathematical Association of America.

Lobastova, N., and M. Hirst (2006). World's top maths genius jobless and living with mother. *Daily Telegraph.* August 20.

Lohne, J. A. (1972). Harriot, Thomas. In C. C. Gillispie (ed.), *Dictionary of scientific biography.* Vol. 6, 124–29. New York: Charles Scribner's Sons.

———. (1979). Essays on Thomas Harriot. I. Billiard balls and laws of collision; II. Ballistic parabolas; III. A survey of Harriot's scientific writings. *Arch. Hist. Exact Sci.* 20(3–4), 189–312.

Machamer, P. (1998). Galileo's machines, his mathematics, and his experiments. In P. Machamer (ed.), *The Cambridge Companion to Galileo,* Cambridge Companions to Philosophy, 53–79. New York: Cambridge University Press.

Mackenzie, D. (2006). Breakthrough of the year: The Poincaré conjecture—proved. *Science* 314(5807), December 22, 1848–49.

MacTutor History of Mathematics Archive, http://www-groups.dcs.st-and.ac.uk/~history. Created by J. J. O'Connor and E. F. Robertson.

Malkevitch, J. (1984). The first proof of Euler's formula. *Mitt. Math. Sem. Giessen* 165, 77–82.

———. (1988). Milestones in the history of polyhedra. In M. Senechal and G. Fleck (eds.), *Shaping space: A polyhedral approach,* proceedings of 1984 conference held in Northampton, MA, 80–92. Boston: Design Science Collection, Birkhäuser Boston.

Martens, R. (2000). *Kepler's philosophy and the new astronomy.* Princeton, NJ: Princeton University Press.

Maurer, S. B. (1983). An interview with Albert W. Tucker. *Two-Year College Mathematics Journal* 14(3), 210–24.

May, K. O. (1965). The origin of the four-color conjecture. *Isis* 56(3), 346–48.

McClellan III, J. E. (1985). *Science reorganized: Scientific societies in the eighteenth century*. New York: A Columbia University Press.

McEwan, I. (1997). *Enduring love*. New York: Anchor Books.

Milnor, J. (1963). *Morse theory*. Based on lecture notes by M. Spivak and R. Wells. *Annals of Mathematics Studies*, no. 51. Princeton, NJ: Princeton University Press.

———. (2003). Towards the Poincaré conjecture and the classification of 3-manifolds. *Notices Amer. Math. Soc.* 50(10), 1226–33.

Möbius, A. F. (1863). Theorie der elementaren Verwandschaften. *Abhandlungen Sächsische Gesellschaft der Wissenschaften* 15, 18–57. Also in *Gesammelte Werke*, vol. 2, Leipzig, 1886, 433–71.

———. (1865). Ueber die Bestimmung des Inhaltes eines Polyëders. *Abhandlungen Sächsische Gesellschaft der Wissenschaften* 17, 31–68. Also in *Gesammelte Werke*, vol. 2, Leipzig, 1886, 473–512.

Morgan, J. W., and G. Tian (2006). Ricci flow and the Poincaré conjecture. http://arXiv:org/abs/math.DG/0607607.

Morse, M. (1929). Singular points of vector fields under general boundary conditions. *Amer. J. Math.* 41, 165–78.

Murasugi, K. (1958). On the genus of the alternating knot. I, II. *J. Math. Soc. Japan* 10, 94–105, 235–48.

———. (1987). Jones polynomials and classical conjectures in knot theory. *Topology* 26(2), 187–94.

Nasar, S., and D. Gruber (2006). Manifold destiny: A legendary problem and the battle over who solved it. *The New Yorker*, August 28, 44–57.

Nash, C. (1999). Topology and physics—a historical essay. In I. M. James (ed.), *History of topology*, 359–415. Amsterdam: North-Holland.

Papakyriakopoulos, C. (1943). A new proof for the invariance of the homology groups of a complex. *Bull. Soc. Math. Grèce* 22, 1–154.

Perelman, G. (2002). The entropy formula for the Ricci flow and its geometric applications. http://arxiv.org/abs/math.DG/0211159.

———. (2003a). Finite extinction time for the solutions to the Ricci flow on certain three-manifolds. http://arxiv.org/abs/math.DG/0307245.

———. (2003b). Ricci flow with surgery on three-manifolds. http://arxiv.org/abs/math.DG/0303109.

Peterson, I. (2003). Recycling topology. *Science News Online* 163(17), April 26.

Pick, G. (1899). Geometrisches zur Zahlenlehre. *Sitzungber. Lotos, Naturwissen Zeitschrift, Prague* 19, 311–19.

Plato (1921). *Theaetetus. Sophist*. With an English translation by H. N. Fowler. New York: G. P. Putnam's Sons.

———. (1972). *Philebus and Epinomis*. Translation and introduction by A. E. Taylor. London: Dawsons of Pall Mall.

———. (2000). *Timaeus*. Translated with an introduction by Donald J. Zeyl. Indianapolis: Hacket Publishing.

Poincaré, H. (1881). Mémoire sur les courbes définiés par une équation differentielle. *J. de Math.* 7, 375–422.

———. (1885). Sur les courbes défines par les équations differentielles. *Journal de mathématiques* 1(4), 167–244.

————. (1895). Analysis situs. *J. Ec. Polytech ser.* 2 1, 1–123.

————. (1899). Complément à l'analysis situs. *Rend. Circ. Math. D. Palermo* 13, 285–343.

————. (1900). Second complément à l'analysis situs. *Proc. Lond. Math. Soc.* 32, 277–308.

————. (1902a). Sur certaines surfaces algèbriques; troisième complément à l'analysis situs. *Bull. Soc. Math. France* 30, 49–70.

————. (1902b). Sur les cycles algèbriques; quatrième complément à l'analysis situs. *J. de Math.* 8, 169–214.

————. (1904). Cinquième complément à l'analysis situs. *Rend. Circ. Math. D. Palermo* 18, 45–110.

————. (1913). *The foundations of science: Science and hypothesis, the value of science, science and method.* Science and Education. New York: The Science Press.

Poinsot, L. (1810). Mémoire sur les polygones et les polyèdres. *Journal de l'école polytéchnique* 4, 16–48.

Pólya, G. (1954). *Induction and analogy in mathematics.* Vol. 1 of *Mathematics and plausible reasoning.* Princeton, NJ: Princeton University Press.

Pont, J.-C. (1974). *La topologie algèbrique des origines à Poincaré.* Paris: Presses Universitaires de France.

Przytycki, J. (1992). A history of knot theory from Vandermonde to Jones. *Aportaciones Matemáticas Comunicaciones* 11, 173–85.

Radó, T. (1925). Über den Begriff von Riemannsche fläche. *Acta Univ. Szeged* 2, 101–20.

Ranicki, A. A., A. J. Casson, D. P. Sullivan, M. A. Armstrong, C. P. Rourke, and G. E. Cooke (1996). *The Hauptvermutung book,* volume 1 of *K-Monographs in Mathematics. A collection of papers of the topology of manifolds.* Dordrecht: Kluwer Academic Publishers.

Read, J. (1966). *Prelude to chemistry: An outline of alchemy, its literature and relationships.* Cambridge, MA: The M.I.T. Press.

Riasanovsky, N. V. (1993). *A History of Russia* (5th ed.). New York: Oxford University Press.

Richeson, D. (2007). The polyhedral formula. In R. Bradley and E. Sandifer (eds.), *Leonhard Euler: Life, work and legacy.* Vol. 5 of *Studies in the history and philosophy of mathematics,* 421–39. Amsterdam: Elsevier.

Riemann, G. F. B. (1851). *Grundlagen für eine allgemeine Theorie der Functionen einer veränderlichen complexen Grösse.* PhD thesis, Göttingen.

————. (1857). Theorie der Abel'schen Functionen. *Journal für Mathematik* 54, 101–55. Also in *Gesammelte Mathematische Werke und Wissenschaftlicher Nachlass,* Berlin: Springer, 1990, 88–142.

Russell, B. (1957). The study of mathematics. In *Mysticism and Logic,* 55–69. Garden City, NY: Doubleday.

————. (1967). *The autobiography of Bertrand Russell,* vol. 1. Boston: Little, Brown.

Sachs, H., M. Stiebitz, and R. J. Wilson (1988). An historical note: Euler's Königsberg letters. *Journal of Graph Theory* 12(1), 133–39.

Salzberg, H. W. (1991). *From caveman to chemist: Circumstances and achievements.* Washington DC: American Chemical Society.

306   REFERENCES

Samelson, H. (1995). Descartes and differential geometry. In *Geometry, topology, & physics,* Conf. Proc. Lecture Notes in Geometry and Topology, IV, 323–28. Cambridge, MA: Internat. Press.

———. (1996). In defense of Euler. *Enseign. Math.* (2) 42(3–4), 377–82.

Sandifer, E. (2004). How Euler did it: V, E and F, parts 1 and 2. Mathematical Association of America Online. http://www.maa.org/news/howeulerdidit.html.

Sarkaria, K. S. (1999). The topological work of Henri Poincaré. In *History of topology,* 123–67. Amsterdam: North-Holland.

Schechter, B. (1998). *My brain is open: The mathematical journeys of Paul Erdős.* New York: Touchstone.

Schläfli, L. (1901). Theorie der vielfachen Kontinuität. *Denkschr. Schweiz. naturf. Ges.* 38, 1–237.

Scholz, E. (1999). The concept of manifold, 1850–1950. In I. M. James (ed.), *History of topology,* 25–64. Amsterdam: North-Holland.

Seifert, H. (1934). Über das Geschlecht von Knotten. *Math. Ann.* 110, 571–92.

Seifert, H., and W. Threlfall (1980). *Seifert and Threlfall: A textbook of topology,* vol. 89 of *Pure and Applied Mathematics.* Translated from the German edition of 1934 by Michael A. Goldman, with a preface by Joan S. Birman. With "Topology of 3-dimensional fibered spaces" by Seifert, translated from the German by Wolfgang Heil. New York: Academic Press (Harcourt Brace Jovanovich Publishers).

Senechal, M. (1988). A visit to the polyhedron kingdom. In M. Senechal and G. Fleck (eds.), *Shaping space: A polyhedral approach,* proceedings of 1984 conference held in Northampton, MA, 3–43. Boston, Design Science Collection, Birkhäuser Boston.

Shakespeare, W. (1992). *Hamlet.* New York: Dover.

———. (2002). *Twelfth night.* Woodbury, CT: Barron's Educational Series.

Simmons, G. F. (1992). *Calculus gems: Brief lives and memorable mathematics.* With portraits by Maceo Mitchell. New York: McGraw-Hill.

Simpson, J., and E. Weiner (eds.) (1989). *Oxford English Dictionary* (2nd ed.). Oxford: Clarendon Press.

Sloane, N. J. A. (2007). The online encyclopedia of integer sequences. http://www.research.att.com/~njas/sequences.

Smale, S. (1961). Generalized Poincaré's conjecture in dimensions greater than four. *Ann. of Math.* (2) 74, 391–406.

———. (1990). The story of the higher dimensional Poincaré conjecture (what really actually happened on the beaches of Rio). *Math. Intelligencer* 12(2), 44–51.

———. (1998). Mathematical problems for the next century. *Math. Intelligencer* 20(2), 7–15.

Sommerville, D. M. Y. (1958). *An introduction to the geometry of n dimensions.* New York: Dover.

Speziali, P. (1973). L'huillier, Simon-Antoine-Jean. In C. C. Gillispie (ed.), *Dictionary of scientific biography.* Vol. 8, 305–7. New York: Charles Scribner's Sons.

Stallings, J. (1960). Polyhedral homotopy-spheres. *Bull. Amer. Math. Soc.* 66, 485–88.

Stallings, J. (1962). The piecewise-linear structure of Euclidean space. *Proc. Cambridge Philos. Soc.* 58, 481–88.

Steiner, J. (1826). Leichter Beweis eines stereometrischen Satzes von Euler. *Journal für die reine und angewandte Mathematik* 1, 364–67.

Steinitz, E. (1922). Polyeder und Raumeinteilungen. In W. F. Meyer and H. Mohrmann (eds.), *Encyclopädie der mathematischen Wissenschaften*. Vol. 3 (Geometrie), 1–139. Leipzig: Teubner.

Stillwell, J. (2002). *Mathematics and its history* (2nd ed.). Undergraduate Texts in Mathematics. New York: Springer-Verlag.

Struik, D. J. (1972). Gergonne, Joseph Diaz. In C. C. Gillispie (ed.), *Dictionary of scientific biography*. Vol. 5, 367–69. New York: Charles Scribner's Sons.

Tait, P. G. (1883). Johann Benedict Listing. *Nature* 28, February 1, 316. Also in *Scientific Papers of Peter Guthrie Tate*, vol. 2, Cambridge: Cambridge University Press, 81–84.

———. (1884). Listing's *Topologie*. Introductory address to the Edinburgh Mathematical Society, November 9, 1883. *Philosophical Magazine* 17(5), January, 30–46.

Taubes, G. (1987). What happens when hubris meets nemesis. *Discover* 8, July, 66–77.

Taylor, A. E. (1929). *Plato: The man and his work*. New York: The Dial Press.

———. (1962). *A commentary on Plato's Timaeus*. London: Oxford University Press.

Terquem, O. (1849). Sur les polygones et les polyèdres étoilés, polygones funiculaires. *Nouv. Ann. Math.* 8, 68–74.

Terrall, M. (1990). The culture of science in Frederick the Great's Berlin. *Hist. Sci.* 28, 333–64.

Thistlethwaite, M. B. (1987). A spanning tree expansion of the Jones polynomial. *Topology* 26(3), 297–309.

Thomassen, C. (1992). The Jordan-Schönflies theorem and the classification of surfaces. *Amer. Math. Monthly* 99(2), 116–30.

Thoreau, H. D. (1894). In F. B. Sanborn (ed.), *Familiar Letters of Henry David Thoreau*. Boston: Houghton, Mifflin and Co.

Thurston, W. P. (1982). Three-dimensional manifolds, Kleinian groups and hyperbolic geometry. *Bull. Amer. Math. Soc. (N.S.)* 6(3), 357–81.

———. (1997). *Three-dimensional geometry and topology*, vol. 1. Princeton, NJ: Princeton Univ. Press.

Tucker, A. W., and F. Nebeker (1990). Lefschetz, Solomon. In C. C. Gillispie (ed.), *Dictionary of scientific biography*. vol. 18, 534–39. New York: Charles Scribner's Sons.

Turnbull, H. W. (1961). *The great mathematicians*. New York: New York University Press.

Twain, M. (1894). *Tom Sawyer abroad*. New York: Jenkins & Mccowan.

van der Waerden, B. L. (1954). *Science awakening*. English translation by Arnold Dresden. Groningen, Netherlands: P. Noordhoff.

Vanden Eynde, R. (1999). Development of the concept of homotopy. In *History of topology*, 65–102. Amsterdam: North-Holland.

Vandermonde, A.-T. (1771). Remarques sur les problèmes de situation. *Mémoires de l'Académie Royale des Sciences de Paris* 15, 566–74.

Varberg, D. E. (1985). Pick's theorem revisited. *Amer. Math. Monthly* 92(8), 584–87.

von Fritz, K. (1975). Pythagoras of Samos. In C. C. Gillispie (ed.), *Dictionary of scientific biography*. Vol. 11, 219–25. New York: Charles Scribner's Sons.

von Staudt, K. G. C. (1847). *Geometrie der Lage*. Nürnberg: Bauer und Raspe.

Vucinich, A. (1963). *Science in Russian culture: A history to 1860*. Stanford, CA: Stanford University Press.

Waterhouse, W. C. (1972). The discovery of the regular solids. *Arch. Hist. Exact Sci.* 9, 212–21.

Weeks, J. R. (2002). *The shape of space,* 2nd ed. New York: Marcel Dekker.

Weibel, C. A. (1999). History of homological algebra. In *History of topology,* 797–836. Amsterdam: North-Holland.

Weil, A. (1984). Euler. *Amer. Math. Monthly* 91(9), 537–42.

Wells, D. (1990). Are these the most beautiful? *Math. Intelligencer* 12(3), 37–41.

Weyl, H. (1989). *Symmetry.* Reprint of the 1952 original. Princeton Science Library. Princeton, NJ: Princeton University Press.

Wilson, R. J. (1986). An Eulerian trail through Königsberg. *Journal of Graph Theory* 10(3), 265–75.

———. (2002). *Four colors suffice: How the map problem was solved.* Princeton, NJ: Princeton University Press.

Youschkevitch, A. P. (1971). Euler, Leonhard. In C. C. Gillispie (ed.), *Dictionary of scientific biography.* Vol. 4, 467–84. New York: Charles Scribner's Sons.

Zeeman, E. C. (1961). The generalised Poincaré conjecture. *Bull. Amer. Math. Soc.* 67, 270.

———. (1962). The Poincaré conjecture for $n \geq 5$. In *Topology of 3-manifolds and related topics (Proc. The Univ. of Georgia Institute, 1961),* 198–204. Englewood Cliffs, N.J.: Prentice-Hall.

ILLUSTRATION CREDITS

Fig. 1.3	Courtesy of Archives and Special Collections, Dickinson College, Carlisle, PA
Fig. 4.1	Courtesy of Smithsonian Institution Libraries, Washington, DC
Fig. 6.3 (top left)	Courtesy of Peter Cromwell
Fig. 8.2	Courtesy of Mark Richeson
Fig. 9.1	Library of Congress, Prints and Photographs Division, LC-USZ62-61365
Fig. 10.1	Courtesy of the Bibliothèque de l'Institut de France, Paris. Photo credit: Réunion des Musées Nationaux/Art Resource, NY
Fig. 10.13	Courtesy of Smithsonian Institution Libraries, Washington, DC
Fig. 16.7 (left)	M.C. Escher's "Moebius Strip II" © 2007 The M.C. Escher Company-Holland. All rights reserved. www.mcescher.com
Fig. 18.3	Library of Congress, Prints and Photographs Division, LC-USZ62-64292
Fig. 18.5	Courtesy of the Archives of the Mathematisches Forschungsinstitut Oberwolfach
Fig. 19.12	Courtesy of Smithsonian Institution Libraries, Washington, DC
Fig. 19.13	Courtesy of the Archives of the Mathematisches Forschungsinstitut Oberwolfach
Fig. 23.8	Courtesy of the Archives of the Mathematisches Forschungsinstitut Oberwolfach

INDEX